PLANETARY MOTIONS

Recent Titles in
Greenwood Guides to Great Ideas in Science
Brian Baigrie, Series Editor

Electricity and Magnetism: A Historical Perspective
Brian Baigrie

Evolution: A Historical Perspective
Bryson Brown

The Chemical Element: A Historical Perspective
Andrew Ede

The Gene: A Historical Perspective
Ted Everson

The Cosmos: A Historical Perspective
Craig G. Fraser

Heat and Thermodynamics: A Historical Perspective
Christopher J. T. Lewis

Earth Cycles: A Historical Perspective
David Oldroyd

Quantum Mechanics: A Historical Perspective
Kent A. Peacock

Forces in Physics: A Historical Perspective
Steven Shore

PLANETARY MOTIONS

A Historical Perspective

Norriss S. Hetherington

Greenwood Guides to Great Ideas in Science
Brian Baigrie, Series Editor

GREENWOOD PRESS
Westport, Connecticut • London

Library of Congress Cataloging-in-Publication Data

Hetherington, Norriss S., 1942–
    Planetary motions: a historical perspective / Norriss S. Hetherington.
        p.   cm.—(Greenwood guides to great ideas in science, ISSN: 1559–5374)
    Includes bibliographical references and index.
    ISBN 0–313–33241–X (alk. paper)
    1. Astronomy—History. 2. Astronomy—Mathematics—History. 3. Planetology.
4. Science—History. I. Title. II. Series.
    QB15.H59 2006
    520.9—dc22   2006005043

British Library Cataloguing in Publication data is available.

Library of Congress Catalog Card Number: 2006005043
ISBN: 0–313–33241–X
ISSN: 1559–5374

First published in 2006

Greenwood Press, 88 Post Road West, Westport, CT 06881
An imprint of Greenwood Publishing Group Inc.
www.greenwood.com

Printed in the United States of America

∞™

The paper used in this book complies with the
Permanent Paper Standard issued by the National
Information Standards Organization (Z39.48–1984).

10 9 8 7 6 5 4 3 2 1

for
Arael, Michael, and Evan
With the hope
that my grandchildren
will come to appreciate
the history of planetary astronomy
from Plato to Copernicus and Newton,
and how changes in science
rippled outward
through Western civilization
and changed their world for the better.

CONTENTS

SERIES FOREWORD

The volumes in this series are devoted to concepts that are fundamental to different branches of the natural sciences—the gene, the quantum, geological cycles, planetary motion, evolution, the cosmos, and forces in nature, to name just a few. Although these volumes focus on the historical development of scientific ideas, the underlying hope of this series is that the reader will gain a deeper understanding of the process and spirit of scientific practice. In particular, in an age in which students and the public have been caught up in debates about controversial scientific ideas, it is hoped that readers of these volumes will better appreciate the provisional character of scientific truths by discovering the manner in which these truths were established.

The history of science as a distinctive field of inquiry can be traced to the early seventeenth century when scientists began to compose histories of their own fields. As early as 1601, the astronomer and mathematician, Johannes Kepler, composed a rich account of the use of hypotheses in astronomy. During the ensuing three centuries, these histories were increasingly integrated into elementary textbooks, the chief purpose of which was to pinpoint the dates of discoveries as a way of stamping out all too frequent propriety disputes, and to highlight the errors of predecessors and contemporaries. Indeed, historical introductions in scientific textbooks continued to be common well into the twentieth century. Scientists also increasingly wrote histories of their disciplines—separate from those that appeared in textbooks—to explain to a broad popular audience the basic concepts of their science.

The history of science remained under the auspices of scientists until the establishment of the field as a distinct professional activity in middle of the twentieth century. As academic historians assumed control of history of science writing, they expended enormous energies in the attempt to forge a distinct and autonomous discipline. The result of this struggle to position the history of science as an intellectual endeavor that was valuable in its own right, and not merely in consequence of its ties to science, was that historical studies of the natural sciences were no longer composed with an eye toward educating a wide audience that included non-scientists, but instead were composed with the aim of being consumed by other professional historians of science. And as historical breadth was sacrificed for technical detail, the literature became increasingly daunting in its technical detail. While this scholarly work increased our understanding of the nature of science, the technical demands imposed on the reader had the unfortunate consequence of leaving behind the general reader.

As Series Editor, my ambition for these volumes is that they will combine the best of these two types of writing about the history of science. In step with the general introductions that we associate with historical writing by scientists, the purpose of these volumes is educational—they have been authored with the aim of making these concepts accessible to students—high school, college, and university—and to the general public. However, the scholars who have written these volumes are not only able to impart genuine enthusiasm for the science discussed in the volumes of this series, they can use the research and analytic skills that are the staples of any professional historian and philosopher of science to trace the development of these fundamental concepts. My hope is that a reader of these volumes will share some of the excitement of these scholars—for both science, and its history.

Brian Baigrie
University of Toronto
Series Editor

INTRODUCTION

Historians provide a "usable past": lessons from history that spare readers otherwise necessary and often unpleasant hard learning from experience. Learning is the foundation for good civic judgment and a better-governed society.

Historians also offer new and interesting ideas valuable for themselves and for whatever feelings of pleasure such knowledge may evoke in readers, irrespective of any moral or civic lessons to be drawn from history. This pleasure follows from a change in brain chemistry; it is perhaps not dissimilar to the effects of drugs, licit and otherwise.

There is little need to elaborate on or argue for the recreational use of history. Readers abusing this substance for pleasure already are too addicted to quit. That leaves to be made a case for a usable past, and, for this particular book, a plausible argument that insight into developments in planetary astronomy, beginning with the ancient Greeks and culminating in the Copernican and Newtonian revolutions, might somehow have relevance for contemporary problems. Here goes . . .

A pressing problem of the moment, and likely for years to come, is Islamic fundamentalism and its conflict with Western civilization: the clash of ideologies, cultures, and religious beliefs. The West is truly a superpower, and the only power on planet Earth strong enough to destroy itself. Terrorism, in contrast, must provoke fear and overreaction if it is to achieve its ends. To survive the long reach of the Islamist challenge and the threat it poses to Western civilization, we must know what differentiates the West from the rest. We must know ourselves as thoroughly as we know our competition and our enemies.

Two millennia of planetary astronomy, leading up to the scientific revolution of the sixteenth and seventeenth centuries, are rooted in Plato, particularly in his book the *Republic*, in passages on the proper education of the philosopher-king. Contemporary Islamic fundamentalism is also rooted in this early stage of Western civilization, in Plato's ideal of a philosopher-king ruling wisely over society. Iran's Khomeini was a devotee of Plato, and the ayatollah's political philosophy was rule of the jurisprudent: a state ruled over by a theocratic philosopher-king. Only this one right-minded man, most learned in Islamic law for Khomeini and most learned in philosophy for Plato, could guide his country correctly.

Plato's way of thinking in various intellectual realms, including religion, politics, and science, was replaced in the West in the aftermath of the Copernican and Newtonian revolutions. That Plato's way of thinking still reigns in Islamic fundamentalism is at the root of the conflict between what Western civilization was thousands of years ago and what it is now.

Before the scientific revolution, there were great civilizations in ancient China, Greece, and Rome, and—closer to our time—in tenth-century Cairo, in thirteenth-century Tehran, and in sixteenth-century Istanbul. Then Muslim civilization was more advanced than its Christian counterpart. The ensuing political and cultural advance in the West, which overturned that ranking, coincided with the scientific revolution and owes much to it. Terrorist attacks on Western values are fed by a sense of humiliation over this reversal of standing, a feeling of impotence, and the resulting rage.

Modern science and technology are the foundation of the material wealth and power of Western civilization and also lie at the heart of the ideas and ideals that have come to differentiate the West from the rest. At the heart of the scientific revolution in the West were changes wrought by Nicholaus Copernicus and Isaac Newton in understanding planetary motions.

The Copernican revolution saw the demise of the finite, closed, and hierarchically ordered universe of medieval belief. It was replaced by an indefinite or even infinite universe consisting of components and laws but lacking value concepts: perfection, harmony, meaning, and purpose. No longer was the universe specifically created for humankind.

The Newtonian revolution saw the replacement of God's rule with a purely physical theory and the separation of science and religion, previously joined in Western thought. In the eighteenth-century Enlightenment, critical human reason freed people from ignorance, from prejudices, and from unexamined authority, including religion and the state. Newtonian planetary astronomy proved the amazing ability of human reason. Newton's success encouraged others to apply reason in other realms. Political thinkers now had confidence that they could determine the natural laws governing human association, and the American and French revolutions followed. Adam Smith attempted to discover general laws of economics.

Muslim political and cultural stagnation, and consequently relative failure during a period of rapid Western advance, occurred at least in part because that culture did not embrace the Copernican and Newtonian revolutions. Many civilizations have stagnated because reason lacked independence from religion and also the status necessary to challenge received ideas. The separation of reason from religion, its enhanced status, and the consequent subjection of government to reasoned analysis is one of the West's most valuable legacies from the Newtonian revolution.

Secular modernity now underlies the economic, political, social, and cultural world of the West, in stark contrast to the fundamentalist Muslim world. Its traditional values are increasingly unable to compete with Western values but are yet to be significantly modified or replaced. Islamic fundamentalism would return all the world, including infidels, once again under the rule of true believers, to a religious society under their Allah's laws.

Western civilization is not an inevitable and unending happy state of affairs that other cultures will automatically achieve in time. Western civilization may not survive without a deep and widespread understanding of how it came about and the conditions necessary for its flourishing. We must understand our history if we are to defend our freedom. History does matter.

LIST OF ILLUSTRATIONS

AN INTRODUCTION TO THE HISTORY OF SCIENCE

Rather than plunging immediately into an account of planetary motions and the Copernican and Newtonian revolutions, this first chapter looks at the history of science in general and provides a framework within which historical facts might be arranged and comprehended.

There are, to be sure, historical facts. But different historians find significance in different combinations of facts. They weave from different threads, or even from the same threads, different patterns. Different historians imagine different periodizations and create different characterizations of scientific activity over the ages. In addition to decribing historical change, historians of science also attempt to identify causes and to weigh the relative importance of various events.

Too often, history has been written as merely a record of events and a chronological narrative. The history of science then becomes a tale of great discoveries and important laws arranged in the order they were discovered.

Implicit in this kind of history is the assumption that science is primarily a collection of facts. The history of science, however, is more a creation of the human imagination than a cumulative list of names and discoveries and dates. The history of science is more an ongoing activity than a lifeless collection of old theories and observations.

Scientist-historians writing chronologies of great discoveries are said to be afflicted with the imaginary disease *precursitis*. They unthinkingly assume that ancient scientists were working on the same problems and using the same methods as modern scientists are. Hence the search for ancient precursors of the observations and theories now acclaimed in textbooks. The myopic result is a chronology of accumulating achievement. Scientists throughout history are seen to have contributed small bits, and occasionally large chunks, to an ever-growing pile.

The ideal of cumulative, systematized positive knowledge is psychologically comforting. It was especially so in the immediate aftermath of World War I and the destruction of European civilization. Only science among all human activities survived that war with its good reputation intact. Even music, in which Germany excelled over all other nations, was found wanting for having failed to civilize its practitioners. Science was praised as the only truly cumulative and progressive human activity. The history of science was believed to be the only history illustrating the progress of humankind.

Instead of a cumulative result, however, science can be thought of as a process. History of science then focuses on scientists thinking about and wrestling with problems.

Gerald Holton, both a working scientist and a historian of science, has proposed nine different facets of science for study by historians:

1. So-called scientific facts, data, laws, theories, and techniques that the scientist was aware of.
2. The conceptual development of shared scientific beliefs preceding the new discovery.
3. The creative insight that guided the scientist. Einstein wrote that scientific hypotheses are free inventions of the human intellect.
4. The scientist's own personal struggle.
5. Possible connections between a person's scientific work and other aspects of his or her life.
6. The sociological setting, including interactions with other scientists, funding for research, and the prestige society accords to scientists.
7. Cultural developments influencing science and influenced by science, including connections between technology, science, and society; between science and ethics; and between science and literature.
8. A logical analysis of the scientific work, although a rational reconstruction may differ from what actually happened, especially in the creative phase of the invention or discovery of a scientific theory.
9. An analysis of *themata* constraining or motivating scientists and guiding or polarizing scientific communities. Often people are not consciously aware of all their underlying beliefs, values, and worldviews. These themata affect quasi-aesthetic choices of scientists.

Themata, though last on Holton's list, are first in his heart. Themata can exert a strong grip on a scientist or on a scientific community. A historian of science doing a thematic analysis is like a folklorist or anthropologist looking for and identifying recurring general themes in the preoccupations of individuals and of a society. For example, a belief still in force in Copernicus's time was that celestial motions were circular and uniform.

Holton's *themata* are similar to Thomas Kuhn's *paradigms:* universally recognized achievements providing model problems and solutions. Kuhn, a historian and philosopher of science, and trained as a scientist in theoretical physics, brought together his ideas about science in a book, *The Structure of Scientific Revolutions* (1962). More than a million copies have been printed, in more than a dozen languages. It is the most influential book ever written on how science works.

Kuhn objected that science textbooks present science as a collection of facts, theories, and methods, and scientists as striving successfully to contribute a small bit to the ever-growing pile. The history of science in textbooks is linear and cumulative. Scientists of earlier ages are imagined working on the same problems and using the same methods as modern scientists do.

Such history of science leaves readers blind to everything but the current scientific perspective, blind especially to different ways of doing science in the past. A Kuhnian history of science, on the other hand, reveals unexpected discoveries subverting scientific traditions and leading scientists to new beliefs and new ways of doing science.

The historian of science Steve Brush (1974) has asked, tongue only partially in cheek, if it is safe to expose students to Kuhn's ideas. Will knowing that current scientific beliefs eventually will be overthrown and abandoned discourage students from studying current science? Should the history of science be X-rated?

Opponents of science conclude, from Kuhn's demonstration that personalities and politics play a role in science, that science and scientists are subjective and irrational. Not surprisingly, scientists object vehemently to such a characterization of themselves and their profession.

Kuhn's (1962) basic outline of how science works is summarized in five terms:

1. *Normal science* is the continuation of a research tradition. It seeks facts shown by theory to be of interest. The result is anticipated.
2. *Paradigm* is a universally recognized achievement temporarily providing model problems and solutions to a community of practitioners. Normal science is research based on a paradigm. Paradigms tell scientists about the entities that nature contains and about the ways in which these entities behave.
3. *Anomaly* is a violation of expectation. It is a discovery for which an investigator's paradigm has not prepared him or her. Normal science does not aim at novelties of fact or theory, but it does produce them.
4. *Crisis* occurs when an anomaly is judged worthy of concerted scrutiny yet continues to resist increased attention. A crisis follows repeated failures to make an anomaly conform. It demands large-scale paradigm destruction and major shifts in the problems and techniques of normal science. External social, economic, and intellectual conditions may also help transform an anomaly into a crisis.
5. *Scientific revolution* is an extraordinary episode. Scientific beliefs, values, and worldviews are abandoned. Ruling paradigms are replaced by incompatible new paradigms. Paradigm choice cannot be settled by logic and experiment. Neither political nor scientific revolutions can be resolved within existing rules. Revolutionary differences inevitably end in attempts at mass persuasion, often including force.

Kuhn's picture of science is evolutionary. It is not, however, evolution toward something in particular. It is not toward what we might think we want to know, nor toward some ultimately true understanding of the universe. Rather, it is evolution away from what we do know. Kuhn's understanding of the nature of science denies progress in the sense of a historical movement in a particular desirable direction. Rather, he sees science changing and evolving, but in no particular direction. The denial of scientific progress toward an ultimate truth may be the most revolutionary of Kuhn's ideas.

2

BABYLONIAN PLANETARY ASTRONOMY

Gerald Holton urges historians of science to study how scientists think about and wrestle with problems, and Thomas Kuhn directs attention to paradigms and the scientific activity carried out under the guidance of paradigms. Historians of science ask of different civilizations at different times what people investigating nature (persons now called scientists) thought they were doing.

Mesopotamian civilizations in the valley between the Tigris and Euphrates rivers (modern-day Iraq) produced some of the earliest written documentation of astronomical activity concerning the motions of the planets, the Sun, and the Moon. Babylonian astronomy, as it is more usually called, was preserved on clay tablets. Writers pressed a sharpened stylus or stick into soft clay tablets about the size of a hand, leaving wedge-shaped (cuneiform) signs. Tablets were then baked hard.

Explorers in quest of ancient Mesopotamian civilizations at first ignored the clay tablets. Muslims avoided them because the Koran said they were bricks inscribed by demons and baked in hell. Christians thought it blasphemy to revive knowledge of the Assyrians, whom the Lord had determined to destroy along with all record of their wickedness.

In the nineteenth century, however, a French official in Mesopotamia grew curious about some large pieces of sculpture and began excavations. He discovered the ancient city of Nineveh, capital of the Assyrian Empire. The Assyrians had pursued a deliberate policy of terrorism, covering the walls of conquered cities with the skins of residents. Their empire lasted from the ninth century B.C. until a revolution that brought about the destruction of Nineveh in 612 B.C.

The same curious French official next found remains of an earlier empire. The Sumerians controlled the lower part of the Tigris-Euphrates Valley from around 3000 to 2000 B.C.

Existing in the time between the Sumerians and the Assyrians had been the Amorites. Their great leader Hammurabi, famous for his code of laws, ruled from about 1792 to 1750 B.C. Many recovered mathematical texts are from this period.

Most astronomical tablets are from a later period. The successful rebels against the Assyrians, the Chaldeans, ruled for less than a century. Then the Persians captured Babylon in 539 B.C. Next, Alexander the Great conquered the Persian Empire in 330 B.C. His followers ruled Babylon for the last three centuries B.C., and they produced numerous astronomical tablets.

Shortly after the discovery of the ruins of Nineveh, the library of Ashurbanipal (an Assyrian king who died in 626 B.C.) was uncovered. Some 22,000 clay tablets from that library are now stored in the British Museum in London, along with 50,000 clay tablets from the temple library of Nippur, written between 3000 and 450 B.C.

Glamour and Funding: Exploration versus Analysis

Large expeditions digging for clay tablets in deserts are more glamorous than individual scholars sitting at home deciphering the tablets. Consequently, the number of tablets studied is far smaller than the number unearthed.

Similarly, the United States has spent billions of dollars on space exploration, but scientists have looked at only 10 percent of the data brought back and have closely analyzed less than 1 percent. The rest sits in the Washington National Records Center, called by frustrated scientists the "Black Hole."

As many as 500,000 Mesopotamian clay tablets are scattered among the museums of the world, many of them yet to be deciphered. And more lie buried still in Iraq. Enough astronomical texts have been studied, however, to provide at least the beginnings of an understanding of Babylonian astronomy.

A few tablets, in some instances only fragments of tablets, reveal that the Babylonians recorded positions of celestial objects and devised an arithmetical scheme to compute future positions. The tablets probably are from around the time of Hammurabi. Omens, such as the disappearances and reappearances of the planet Venus, were also recorded. Eclipse records have been found from 747 B.C., and systematic observational reports from around 700 B.C.

In the Seleucid period of the last three centuries B.C., the Babylonians devised arithmetic progressions describing motions of celestial bodies. Over 10,000 tablets, apparently from a single ancient archive, reached the British Museum in the 1880s. Among the treasures were over 1,000 texts and fragments of texts relating to astronomy. Of these, about 250 are tablets of celestial positions, and some 70 are parts of procedural texts. The fragments are too incomplete to reveal fully the arithmetic procedures employed by the Babylonians, but the procedures can be inferred or reconstructed from studies of tablets listing positions calculated by means of the procedures.

A table for the years 133–132 B.C., preserved in a clay tablet, illustrates the use of mathematical progressions in calculating the position of the Sun. Such a tabular statement of the positions of a celestial body at regular intervals is now called an *ephemeris*.

Babylonian Ephemeris for the Position of the Sun in the Years 133–132 B.C.

XII	28,55,57,58	22, 8,18,16	Aries
I	28,37,57,58	20,46,16,14	Taurus
II	28,19,57,58	19, 6,14,12	Gemini
III	28,19,21,22	17,25,35,34	Cancer
IV	28,37,21,22	16, 2,56,56	Leo
V	28,55,21,22	14,58,18,18	Virgo
VI	29,13,21,22	14,11,39,40	Libra
VII	29,31,21,22	13,42, 1, 2	Scorpio
VIII	29,49,21,22	13,32,22,24	Sagittarius
IX	29,56,36,38	13,28,59, 2	Capricorn
X	29,38,36,38	13, 7,35,40	Aquarius
XI	29,20,36,38	12,28,12,18	Pices
XII	29, 2,36,38	11,30,48,56	Aries

Source: Otto Neugebauer, *Exact Sciences in Antiquity,* 2nd. ed. (Providence, Rhode Island: Brown University Press, 1957; New York: Dover Publications, 1969), p. 110

A sexagesimal number system (i.e., a system based on the number 60) was used. Positions are reported, for example, as 28, 55, 57, 58, with each succeeding unit representing so many sixtieths of the preceding unit—thus 28 degrees, 55 minutes, 57 seconds, and so forth. Decimal and sexagesimal systems, as well as others, were used concurrently in Mesopotamia. (The sexagesimal system is more cumbersome than the decimal system, but we still divide our hours into 60 minutes and our minutes into 60 seconds.)

The first column of the table lists the months, starting with month XII of the year 133 B.C. Following this are months I through XII for the year 132 B.C.

The second column lists how far the Sun moves in each month. During month I, the Sun moves 28 degrees, 37 minutes, 57 seconds, and $^{58}/_{60}$ second (line 2, column 2). Add this movement to the position of the Sun at the end of the previous month (line 1, column 3), and the result is the position of the Sun at the end of month I (line 2, column 3).

The third column gives the position of the Sun. It is obtained by adding to the initial position the amount of motion during the month, listed in the second column. For example, adding the top line of the third column and the second line of the second column produces the second line in the third column—after subtracting 30 degrees and listing the position 30 degrees ahead in the next segment of the zodiac.

The fourth column is the house of the zodiac in which the Sun resides that month. Each house of the zodiac occupies 30 degrees, one-twelfth of the 360 degrees of a complete circle.

The table shows a decrease of 18 minutes in the distance traveled by the Sun from one month to the next for months I and II of this particular year (132 B.C.).

For months III through VIII, there is an increase of 18 minutes in the distance traversed each month. In the last four months, the distance is 18 minutes less in each consecutive month.

Within each of the three groups of months, the last two sets of numbers in each line of the second column, the solar motion, are unchanged. Obviously, the numbers are not actual observations but have been calculated by adding or subtracting 18 minutes from each preceding line.

> ### Mental Exercise: Calculating from an Ephemeris
>
> Using the translated Babylonian ephemeris for the position of the Sun in the years 133–132 B.C., do the calculations to find the position of the Sun at the end of month V for the year 132 B.C.

The decreasing, increasing, and again decreasing sequence of numbers in the second column (the Sun's motion) is now called a "zigzag" function, reflecting its appearance on a graph. Babylonian astronomers also used a system in which the solar velocity remained constant for several months, after which the Sun proceeded with a different constant speed for several more months before finally reverting to the initial velocity and remaining at that speed for several more months. Graphed, the motion looks like a series of steps up and down, and it is called a "step" function. *Zigzag* and *step* are useful labels but potentially misleading. Babylonians are not known to have used graphs.

Nor do we have any evidence that Babylonians constructed geometrical models of the motions of celestial bodies. Nor did they express concern about the causes of the motions, at least not in the tablets found and studied so far. Nor did they express any curiosity about the physical composition of the celestial bodies. The goal of ancient Babylonian astronomy seems to have been to predict astronomical appearances but not to make sense of them. Predicting positions did not require any theoretical ideas about the celestial bodies.

Some historians of science celebrate the high level of mathematical theory in late Babylonian astronomy as the first appearance of modern science. The historian of ancient science Otto Neugebauer concluded that Babylonian astronomical texts of the Seleucid period are scientific because everything has been eliminated from the astronomy except observations and the mathematical consequences of an initial hypothesis about the fundamental character of the movements. He admitted only mathematical computation together with empirical observation as the necessary characteristics of science and denied any role to speculative hypotheses of a strongly theoretical nature.

Historians more concerned with underlying motives of science tend to dismiss Babylonian astronomy as little more than a set of mechanical procedures, with no more theoretical content than recipes in a cookbook. Babylonians made astronomical observations and formulated algebraic rules for predicting future positions of the planets, the Sun, and the Moon. They studied *how* the celestial motions went but not *why.* Nor did they seek to develop a single comprehensive mathematical scheme encompassing all their data.

Many historians of science look instead to the Greeks for the birth of modern science. Greek astronomers dealt with planetary motions differently than did the Babylonians, as we will see in following chapters.

3

PLATO AND SAVING THE APPEARANCES

The guiding themata or paradigm of Greek planetary astronomy is attributed to Plato by the philosopher Simplicius of Athens in his commentary on Aristotle's book *On the Heavens*. Around A.D. 500 Simplicius wrote that Plato had set as a task for astronomers to explain the apparently irregular motions of the planets, the Sun, and the Moon as a combination of circular motions with constant speeds of rotation.

To "save the appearances" with a system of uniform circular motions is, in the context of modern science, an arbitrary and absurd task. Granted, the motions of the planets and the Sun and the Moon could be reproduced using, in clever combination, circles of various sizes with unchanging rotational speeds. But it would be a cumbersome contraption. Modern science has achieved a more elegant and informative solution to a more productively formulated problem, at least in the opinion of modern scientists.

The fact remains, however, that a task was set for astronomers; the task was generally accepted; and the task was pursued for nearly two thousand years, from the Greeks in the fourth century B.C. to Copernicus and the European Renaissance in the sixteenth and seventeenth centuries A.D. Historical importance is not necessarily negated by lack of plausibility, especially when plausibility is judged in hindsight by different people in a different age with different standards.

Simplicius lived nearly a thousand years after Plato and Aristotle and the beginning of Greek astronomy, and he lacked direct access to Plato's original writings. Nor is any explicit statement about saving the appearances with a system of uniform circular motions now found in Plato's surviving writings. Hence some historians of science question the central role in the development of planetary theory often assigned to Plato. But whether it began with Plato or slightly afterward, the task for Greek astronomers working in what came to be regarded, rightly or wrongly, as the Platonic tradition, was to save the appearances: to explain apparently irregular motions detected by the senses as a combination of uniform circular motions.

Plato's ideas were taught to pupils at the Academy, possibly the world's first university, which he founded in Athens in about 380 B.C. Plato believed that mathematics provided the finest training for the mind, and over the door of the Academy was written "Let no one unversed in geometry enter here." The Academy survived until banished from Athens by the emperor Justinian in A.D. 529.

Eudoxus, the greatest genius in mathematics and astronomy of his time, may have attended Plato's lectures at the Academy, and certainly he was familiar with Plato's ideas. Upon Eudoxus's report of what Plato said a string of statements followed. Eudoxus's report is lost. However, it was summarized by Eudemus in his own *History of Astronomy*. This work, too, is lost. But it was commented on by Sosigenes in the second century A.D. Sosigenes' work is also lost. It was used, however, by Simplicius, with whom the string of lost-but-summarized works finally ends, in the sixth century A.D. Simplicius wrote:

> Plato lays down the principle that the heavenly bodies' motion is circular, uniform, and constantly regular. Thereupon he sets the mathematicians the following problem: what circular motions, uniform and perfectly regular, are to be admitted as hypotheses so that it might be possible to save the appearances presented by the planets? (Duhem, *To Save the Phenomena*, 5)

Continuing, Simplicius explained:

> The curious problem of astronomers is the following: First, they provide themselves with certain hypotheses: ... Starting from such hypotheses, astronomers then try to show that all the heavenly bodies have a circular and uniform motion, that the irregularities which become manifest when we observe these bodies—their now faster, now slower motion; their moving now forward, now backward; their latitude now southern, now northern; their various stops in one region of the sky; their at one time seemingly greater, and at another time seemingly smaller diameter—that all these things and all things analogous are but appearances and not realities. (Duhem, *To Save the Phenomena*, 23)

A combination of uniform circular motions now seems absurd. Within Plato's philosophy, however, the concept is plausible. And his philosophy is plausible when viewed within the context of his life. Plato's philosophy can be understood as a reaction to the temporary moral values of his age, which left him highly dissatisfied and sent him searching for a new philosophy.

In 479 B.C., a year after the Persians under Xeres I captured and burned Athens, 31 Greek city-states defeated the Persians in decisive land and sea battles. The victory capped 20 years of struggle to stop the westward expansion of the Persian Empire. Now began Greece's Golden Age. Increasingly, Athens dominated. Tribute poured in from other city-states, giving support to

Athenian writers and artists. Led by Sparta, several Greek city-states revolted against Athenian rule, setting off the Peloponnesian Wars of 431–404 B.C. Initially Athens prevailed, but the fortunes of battle shifted after an unsuccessful attack by Athens on Syracuse, a city in Sicily, in 413 B.C. Athens surrendered to Sparta in 404 B.C.

In the turmoil enveloping Greece, searches for a new and more useful philosophy occurred. Socrates, who lived from 469 to 399 B.C., led the reaction against the old philosophy. He was trained as a stonecutter in his father's shop in Athens but preferred to spend his time arguing in the marketplace. There he encouraged the youths of Athens to question every moral precept handed down to them. Elder citizens believed, with justification, that Socrates was demoralizing their children.

Socrates' critical questioning also extended to the government and its actions. In addition to corrupting the youths of Athens, Socrates now was accused of impiety. The charge may have been intended to frighten him into fleeing Athens, but Socrates stayed and forced the issue. He welcomed his trial as a forum for his ideas.

A trial then consisted of two parts. First, guilt or innocence was established. If guilt was established, punishment was determined in a second part. After the jury found Socrates guilty, the prosecution recommended death. Socrates suggested board and lodging at public expense because his actions had been for the public benefit. The jury chose, and Socrates drank the fatal and famous cup of hemlock.

Socrates' death and related political conditions in Athens influenced Plato, who had been Socrates' pupil and a close friend. Born in 427 B.C., Plato was of an age to enter public life at about the time of the defeat of Athens in 404 B.C. Furthermore, both his mother's brother and cousin were members of the oligarchy of the Thirty Tyrants designated by Sparta to rule Athens. In a letter purportedly Plato's and accepted by many, but not all, scholars as genuine, Plato wrote:

> When I was a young man I had the same ambition as many others: I thought of entering public life as soon as I came of age. And certain happenings in public affairs favored me, as follows. The constitution we then had . . . was overthrown; and a new government was set up consisting of . . . thirty other officers with absolute powers. . . . Some of these men happened to be relatives and acquaintances of mine, and they invited me to join them at once. (*Epistles*, VII: 324b–c)

But the actions of the tyrants disgusted Plato. They quelled criticism by intimidation, and opposition by assassination. They met treasury deficits by the arbitrary execution of wealthy individuals for treason, followed by confiscation of the alleged traitors' properties. Also, they attempted to involve Socrates in their illegal actions. Plato chose not to join the government:

> I thought that they were going to lead the city out of the unjust life she had been living and establish her in the path of justice . . . But as I watched they showed in

a short time that the preceding constitution had been a precious thing . . . . I was appalled and drew back from that reign of injustice. (*Epistles*, 325b–c)

A year later a democratic faction drove out the tyrants, and Plato again considered entering politics:

Not long afterwards the rule of the Thirty was overthrown and with it the entire constitution; and once more I felt the desire, though this time less strongly, to take part in public and political affairs. (*Epistles*, 325a–b)

But then the new democracy persecuted Socrates. Plato now determined to set aside political ambition and search for unchanging standards to hold against the shifting judgments of men:

Certain powerful persons brought into court this same friend Socrates, preferring against him a most shameless accusation . . . and the jury condemned and put to death the very man. The more I reflected upon what was happening . . . the more I realized . . . the corruption of our written laws and our customs was proceeding at such amazing speed that whereas at first I had been full of zeal for public life, when I noted these changes and saw how unstable everything was, I became in the end quite dizzy . . . At last I came to the conclusion that all existing states are badly governed and the condition of their laws practically incurable . . . and that the ills of the human race would never end until either those who are sincerely and truly lovers of wisdom come into political power, or the rulers of our cities, by the grace of God, learn true philosophy. (*Epistles*, 325b–326b)

Subsequent experiences confirmed Plato in this opinion. According to legend, in about 388 B.C. the dictator of Syracuse, Dionysius I, asked Plato if he didn't think that he, Dionysius, was a happy man. Plato answered that he thought no one who was not mad would become a tyrant. Enraged, Dionysius supposedly ordered Plato sold into slavery, from which he was rescued by a friend arriving just in time with ransom money. Or maybe on Plato's return voyage to Athens his ship was captured and he was put up for sale in a slave market, where he was ransomed by a friend.

Plato imparted his love of virtue above pleasure and luxury to Dion, the tyrant's brother-in-law. When Dionysius I died, Dion asked Plato to return to Syracuse to help arouse in his nephew, the new tyrant Dionysius II, desire for a life of nobility and virtue. Plato went, though not without trepidation. His worst apprehensions were surpassed:

When I arrived—to make the story short—I found the court of Dionysius full of faction and of malicious reports to the tyrant about Dion. I defended him as well as I could, but was able to do very little; and about the fourth month Dionysius, charging Dion with plotting against the tyranny, had him put aboard a small vessel and exiled in disgrace. Thereupon we friends of Dion were all afraid that one of us might be accused and punished as an accomplice in Dion's conspiracy. About me there even went abroad in Syracuse a report that I had been put to death by Dionysius as the cause of all that had happened. . . . Dionysius . . . devised a means for preventing my departure by bringing me inside the citadel and lodging me there, whence no ship's captain would have dared to take me away . . . Nor would any merchant or guard along the roads leading out of the country have let me pass alone, but would have taken me in charge at once and brought me back to Dionysius . . . I made every effort to persuade Dionysius to let me depart, and we came to an agreement that when peace was restored [Syracuse was then at war with Sicily] and when Dionysius had made his empire more secure, he would recall both Dion and me. . . . On these conditions I promised that I would return. (*Epistles*, 329b–e; 338a)

Peace was restored, and Plato returned to Syracuse. There he found Dionysius II not on fire with philosophy. Indeed, Dionysius II acted as if what Plato said was of no value. Again, Plato sought to leave Syracuse and was held against his will. He became even more deeply entangled in the quarrel between Dionysius II and Dion, which ended in disaster.

Plato's philosophy and its implications for the study of astronomy are particularly understandable as a response to the time of troubles in which he found himself. Reacting to the temporary moral values of his time, Plato searched for unchanging standards. The changing, visible world was without permanent values. So Plato turned to the world of ideas. Here he hoped to find the real and unchanging standards so sadly absent in his world of experience.

In his Allegory of the Cave in his book the *Republic,* Plato explained that the prison of the cave corresponds to the part of the world revealed by the sense of sight. Escape from the cave corresponds to the use of intelligence to reach the real world of knowledge.

In the *Republic,* as in most of his books, Plato created a dialog. He distrusted the fixed, dead words of textbooks and believed that learning could be achieved only through discussion and shared inquiry:

Imagine the condition of men living in a sort of cavernous chamber underground, with an entrance open to the light and a long passage all down the cave. Here they have been from childhood, chained by the leg and also by the neck, so that they cannot move and can see only what is in front of them, because the chains will not let them turn their heads. At some distance higher up is the light of a fire burning behind them; and between the prisoners and the fire is a track with a parapet built along it, like the screen at a puppet-show, which hides the performers while they show their puppets over the top.

I see, said he.

Now behind this parapet imagine persons carrying along various artificial objects, including figures of men and animals in wood or stone or other materials, which project above the parapet. Naturally, some of these persons will be talking, others silent.

It is a strange picture, he said, and a strange set of prisoners.

Like ourselves, I replied; for in the first place prisoners so confined would have seen nothing of themselves or of one another, except the shadows thrown by the fire-light on the wall of the cave facing them, would they?

Not if all their lives they had been prevented from moving their heads.

And they would have seen as little of the objects carried past.

Of course.

Now, if they could talk to one another, would they not suppose that their words referred only to those passing shadows which they saw?

Necessarily.

And suppose their prison had an echo from the wall facing them? When one of the people crossing behind them spoke, they could only suppose that the sound came from the shadow passing before their eyes.

No doubt.

In every way, then, such prisoners would recognize as reality nothing but the shadows of those artificial objects.

Inevitably.

Now . . . suppose one of them set free and forced suddenly to stand up, turn his head, and walk with eyes lifted to the light; all these movements would be painful, and he would be too dazzled to make out the objects whose shadows he had been used to see. What do you think he would say, if someone told him that what he had formerly seen was meaningless illusion, but now, being somewhat nearer to reality and turned towards more real objects, he was getting a truer view? Suppose further that he were shown the various objects being carried by and were made to say, in reply to questions, what each of them was. Would he not be perplexed and believe the objects now shown him to be not so real as what he formerly saw?

Yes.

And suppose someone were to drag him away forcibly up the steep and ragged ascent and not let him go until he had hauled him out into the sunlight, would he not suffer pain and vexation at such treatment, and, when he had come out into the light, find his eyes so full of its radiance that he could not see a single one of the things he was now told were real?

Certainly he would not see them all at once.

Now imagine what would happen if he went down again to take his former seat in the Cave. Coming suddenly out of the sunlight, his eyes would be filled with darkness. He might be required once more to deliver his opinion on those shadows, in competition with the prisoners who had never been released, while his eyesight was still dim and unsteady; and it might take some time to become used to the darkness. They would laugh at him and say that he had gone up only to come back with his sight ruined; it was worth no one's while even to attempt the ascent. If they could lay hands on the man who was trying to set them free and lead them up, they would kill him.

Yes, they would. (*Republic*, VII: 514a–517a)

They did kill Socrates. And students whom Plato tried to free from the shackles of ignorance so they might live enlightened lives in the sunlight of understanding were unappreciative of his effort to teach them.

Plato's concept of reality is plausibly illustrated with a simple example. Think of a circle and draw a circle. Which is real? The circle drawn on paper is not a real circle, no matter how skilled the draftsman. The drawn circle is an imperfect representation in the visible world of experience of a perfect circle. The perfect circle exists only in the mind, only in the world of thought. Plato wrote:

> There is something called a circle. . . . The figure whose extremities are everywhere equally distant from its center is the definition of precisely that to which the names "round," "circumference," and "circle" apply . . . what we draw or rub out, what is turned or destroyed; but the circle itself to which they all refer remains unaffected, because it is different from them. (*Epistles*, VII: 342a–344a)

Taking up the discussion of astronomy in the *Republic*, Plato alluded to its utilitarian benefits: in agriculture, in navigation, and in war. Not for these purposes, however, was astronomy to be esteemed. The true utility of the regimen of study prescribed in the *Republic* was saving souls.

An obvious way of doing astronomy is to observe the motions of the objects in the heavens. But only a discipline dealing with unseen reality will lead the mind upward. The true motions are not to be seen with the eye. It is not by looking at the heavens that one can become truly acquainted with astronomy. Again, Plato put his ideas in a dialog:

Plato's Pervasive Persuasive Philosophy

Plato's philosophy has affected many human intellectual endeavors over many centuries. His emphasis on an ideal reality in place of observed shadows and reflections guided ancient astronomers.

More recently, the twentieth-century Romanian sculptor Constantin Brancusi tried to make his art a working philosophy of Plato. Brancusi wrote that what is real is not the external form, but the essence of things. Starting from this truth it is impossible for anyone to express anything essentially real by imitating its exterior surface. Brancusi looked beneath the surface of human experience for a deeper and truer reality. He abandoned details—forms that express little—keeping only essential anatomical elements. His sculptural kissing couple represent an ideal. They are not just a particular pair of people in love, but all the pairs that lived and loved, here, on Earth. Plato's philosophy paid off big-time: the world auction record (as of 2002) for any sculpture is $18,159,500—for a Brancusi.

The movie *The Matrix* is another modern variation on Plato's cave. We may not be living in the world we perceive. Rather, our brains may be floating in a vat of amniotic fluid and connected to electrodes feeding us sensations, or hallucinations. In *The Matrix*, people lie comatose in cocoons, stacked in incubators, clear pods piled high in high towers. Their brains are penetrated by cables delivering an interactive virtual-reality program, the Matrix, its simulation mistaken for reality.

[Astronomy] is important for military purposes, no less than for agriculture and navigation, to be able to tell accurately the times of the month or year.
I am amused by your evident fear that the public will think you are recommending useless knowledge. . . .

And now, Socrates, I will praise astronomy on your own principles, instead of commending its usefulness in the vulgar spirit for which you upbraided me. Anyone can see that this subject forces the mind to look upwards, away from this world of ours to higher things.

Anyone except me, perhaps, I replied. I do not agree.

Why not?

As it is now handled by those who are trying to lead us upward to philosophy, I think it simply turns the mind's eye downwards.

What do you mean?

You put a too generous construction on the study of "higher things." Apparently you would think a man who threw his head back to contemplate the decorations on a ceiling was using his reason, not his eyes, to gain knowledge. Perhaps you are right and my notion is foolish; but I cannot think of any study as making the mind look upwards, except one which has to do with unseen reality. No one, I should say, can ever gain knowledge of any sensible object by gaping upwards any more than by shutting his eyes and searching for it on the ground, because there can be no knowledge of sensible things. His mind will be looking downwards, though he may pursue his studies lying on his back or floating on the sea.

I deserve to be rebuked. But how did you mean the study of astronomy to be reformed, so as to serve our purposes?

In this way. These intricate traceries in the sky are, no doubt, the loveliest and most perfect of material things, but still part of the visible world, and therefore they fall far short of the true realities—the real relative velocities, in the world of pure number and all perfect geometrical figures, of the movements which carry round the bodies involved in them. These, you will agree, can be conceived by reason and thought, not seen by the eye.

Exactly.

Accordingly, we must use the embroidered heaven as a model to illustrate our study of those realities, just as one might use diagrams exquisitely drawn by some consummate artist like Daedalus. An expert in geometry, meeting with such designs, would admire their finished workmanship, but he would think it absurd to study them in all earnest with the expectation of finding in their proportions the exact ratio of any one number to another.

Of course it would be absurd.

The genuine astronomer, then, will look at the motions of the stars with the same feelings. He will admit that the sky with all that it contains has been framed by its artificer with the highest perfection of which such works are capable. But when it comes to the proportions of day to night, of day and night to month, of month to year, and of the periods of other stars to Sun and Moon and to one another, he will think it absurd to believe that these visible material things go on forever without change or the slightest deviation, and to spend all his pains on trying to find exact truth in them.

Now you say so, I agree.

If we mean, then, to turn the soul's native intelligence to its proper use by a genuine study of astronomy, we shall proceed as we do in geometry, by means of problems, and leave the starry heavens alone. (*Republic*, VII: 527d–530c)

Plato's instruction immediately above, to "leave the starry heavens alone," has dismayed supporters and delighted detractors. The admonishment is anti-empirical, and it could easily lead to a purely speculative study of bodies in motion with no connection to the celestial objects we see. This is especially so if the translation produces in place of "let alone" or "leave" the stronger sense of "dismiss" or "abandon." An injunction to astronomers to dismiss celestial phenomena from the subject matter of their science and to ban sense-perception would result not in a reform of astronomy but in its liquidation.

A few lines earlier, however, Plato instructed that "we must use the embroidered heaven as a model to illustrate our study of those realities." Plato's supporters construe this phrase as scientific, calling for a science of astronomy to ascertain the real motions of the heavenly bodies.

Plato's science, however, cannot be equated with modern science. He wasn't urging astronomers to develop just any theory that would account for observed facts. He was urging astronomers to fit their observations into a predetermined geometrical pattern based on certain a priori assumptions about the behavior of bodies in the sky: that heavenly bodies move in uniform circular orbits.

Plato's science of astronomy had as its subject reality. He wrote that he could not "think of any study as making the mind look upwards, except one which has to do with unseen reality." Plato's reality, however, is not reality as it is now commonly understood. This is evident from the conclusion of the sentence immediately following: "there can be no knowledge of sensible things."

The type of reality discussed in the Allegory of the Cave is what is real for Plato. What is real is the idea or ideal of a perfect geometrical figure, not its imperfect realization in the world of the senses. Plato insisted that the "intricate traceries in the sky are, no doubt, the loveliest and most perfect of material things, but still part of the visible world, and therefore they fall far short of the true realities."

It mattered not to Plato whether a person stared at the ground or at the heavens. As long as he was trying to study any sensible object, he could not be said to have learned anything, because no objects of sense admitted of scientific treatment. In this belief, Plato is antiempirical, and he is antiobservational, but he is not antireality.

Uniform circular orbits have been characterized as "pi in the sky" or "π in the sky," a word play on the fact that the circumference of a circle is equal to π times the diameter. A serious case, however, can be made for the reality of Platonic entities. Plato thought that "forms" are more real than what is observed because logical reasoning is more certain than fallible observations, susceptible as they are to being proved false by subsequent refinements in observational capability. Logical inference and analysis, Plato believed, produce more certain knowledge than does observation.

The aim of ancient Greek geometrical astronomy became to save the appearances with a system of uniform circular motions. The philosophical background for this paradigm is found in Plato's writings. Under the spell of the paradigm, Plato's student Eudoxus would attempt to develop a system of uniform circular motions reproducing the observed motions of the planets.

4

EUDOXUS AND CONCENTRIC SPHERES

Plato encouraged a new approach to astronomy: to devise a combination of uniform circular motions to reproduce the observed motions in the heavens. Whether the Platonic paradigm would die in infancy or grow in strength depended, in part, on the support it received.

Greek society supported playwrights when the citizens of Athens paid to see the productions of Aeschylus, Sophocles, and Euripides; and prizes were awarded at festivals to poets and musicians. No city held geometry in high regard, however, and inquiries in this subject languished, so Plato lamented. Practicing physicians and architects charged fees for their services, and so could philosophers were they able to attract pupils. Alternatives included inherited wealth and the beneficence of a wealthy patron.

The first major attempt to develop the fledgling paradigm of uniform circular motion into a successful science was made by Eudoxus. He arrived in Athens a poor youth, about 23 years old, traveling as assistant to a physician. Diogenes Laertius, a Greek biographer of the third century A.D., wrote in his *Lives of Eminent Philosophers* that when Eudoxus was about 23 years old and in straitened circumstances, he was attracted by the reputation of the Socratics and set sail for Athens with Theomedon the physician who provided for his wants. Diogenes also quoted other writers on Eudoxus, noting that Sotion in his *Successions of the Philosophers* said that Eudoxus was also a pupil of Plato and that Apollodorus stated that Eudoxus flourished about the 103rd Olympiad and died in his fifty-third year.

"Flourishing" often meant one's 40th year. Adding 40 to 368 (the beginning of the 103rd Olympiad) gives 408 B.C. for Eudoxus's birth. Subtracting 23 from 408 gives 385 B.C. as the year of his arrival in Athens. However, Eudoxus's flourishing could have been earlier, in his twenties when already he was famous or later in his life with the development of his planetary theory.

From Athens, Eudoxus traveled to Egypt, where he stayed perhaps for several years and became familiar with priests' astronomical observations. Later he established a school in the Greek city of Cyzicus, and later still, he moved with some of his pupils to Athens.

Unfortunately, nothing Eudoxus wrote has survived. His astronomical system was described briefly by Aristotle in his book *Metaphysics* and, much later (around A.D. 500), by Simplicius in his commentary on Aristotle's book *On the Heavens*. According to Simplicius, Callippus, who had studied Eudoxus's system in Cyzicus, traveled to Athens, where he stayed with Aristotle correcting and completing, with Aristotle's help, the discoveries of Eudoxus. By Simplicius's time, both Eudoxus's and Callippus's books about the planetary system were lost. Simplicius had his information from Sosigenes (around A.D. 250), who had relied on a history of astronomy by Eudemus (a pupil of Aristotle, around 350 B.C.).

In the nineteenth century A.D., Giovanni Schiaparelli, an Italian astronomer more famous—or infamous—for his observations of *canali* on Mars, attempted to reconstruct Eudoxus's system. The reconstruction is compatible with what Aristotle and Simplicius wrote about Eudoxus, but it assumes that Eudoxus's system accounted for many of the astronomical phenomena known in Schiaparelli's time; thus it may attribute to Eudoxus a more advanced knowledge of astronomy than he actually possessed and, consequently, also a more detailed and accurate astronomical model.

A basic observational fact that any astronomical system must account for is the movement of the stars overhead each evening. Eudoxus placed all the stars on one sphere rotating with a uniform speed around the central Earth in 24 hours. This is equivalent to a rotating Earth and a fixed sphere of the stars; the observational consequences of the systems are identical. (Relative motion of a star with respect to another would not be detected for nearly two thousand years and thus did not trouble Eudoxus or complicate his model.)

The apparent motion of the Sun presents a more difficult problem. First, there had to be an outer sphere rotating with a period of 24 hours to produce the apparent daily movement of the Sun across the sky. Again, as with the stars, the outer sphere for the Sun produced the apparent motion now attributed to a rotating Earth.

Eudoxus next would have needed a second sphere rotating with a period of a year and its axis tilted relative to the axis of the outer sphere to move the Sun higher in the sky in summer and lower in winter as well as around the heavens with a period of a year. The axis of this inner sphere would have been fixed to the outer sphere and thus carried around with a 24-hour period.

Eudoxus ignored the changeable velocity of the Sun, already discovered by his time. This decision, whatever its justification, saved him much trouble. Eudoxus did add a third sphere, though, to account for a belief now known to have been mistaken. Continually, it would be a problem to decide which observations were accurate and should be incorporated into a model. Eudoxus

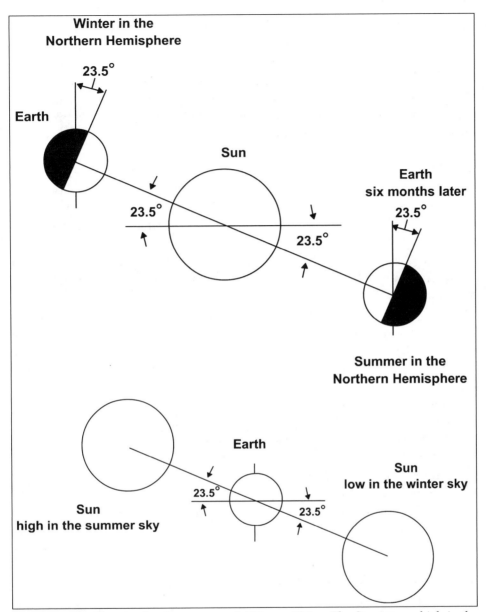

Figure 4.1: The Sun in the Summer and Winter Skies. The Sun is seen high in the summer sky and low in the winter sky because the Earth's axis of rotation is not perpendicular to the plane of the Earth's orbit around the Sun, but is inclined at an angle of about 23.5 degrees. Thus the Sun's apparent annual motion carries it alternately above and below the plane of the Earth's equator.

may not necessarily have intended to account for all observations, however much modern scientists try to.

Eudoxus devised a similar system of spheres for the Moon. First, as with all celestial objects, there was the outer sphere rotating once every 24 hours, producing appearances now attributed to the daily rotation of the Earth.

The Moon circles the Earth approximately once a month. To produce this monthly motion, Eudoxus added a second sphere attached to the first and rotating west to east with a period of one lunar month.

There is also a small variation in the latitude of the Moon. It moves at times slightly above and at other times slightly below an imaginary plane containing the Earth and the Sun. (The modern explanation for this observed phenomenon is that the plane of the Moon's orbit around the Earth is inclined at an angle of approximately five degrees to the plane of the Earth's orbit around the Sun.) Eudoxus added a third sphere, presumably to produce variations in lunar latitude, although we cannot be certain that he knew of this phenomenon. Reversing the order of the middle and inner lunar spheres would bring Eudoxus's model into better agreement with modern observation, and historians should not ignore the possibility that an error occurred somewhere along the way in the transmission of the model. On the other hand, rewriting ancient reports to conform to modern knowledge would be a highly questionable way of doing history.

Eudoxus did not take into account variation in the Moon's speed. Perhaps he was unaware of it, though Callippus certainly knew of this motion about three decades later, around 325 B.C. Alternatively, Eudoxus might have been aware of the phenomenon but chose not to recognize it as requiring a place in his system.

Planetary motions presented a more difficult problem than did the motions of the Sun and the Moon. The planets display retrograde motions: sometimes they cease their motions relative to the stars, turn back temporarily, retrace small parts of their paths, and then change direction once more and resume their voyages around the heavens. Eudoxus's task was to devise a model consisting of uniform circular motions only, yet producing the apparent retrograde motions.

Eudoxus first gave each planet an outer sphere to carry it around the Earth with a period of 24 hours. Second spheres moved the planets around the heavens, with periods of a year for Venus and for Mercury, and longer periods for the outer planets. To produce observed motions in latitude, Eudoxus added third spheres for each planet. So far, the planetary solutions followed the solutions for the Sun and the Moon.

To produce the observed retrograde motions, Eudoxus added a fourth sphere. By a clever combination of inclinations and speeds of revolution of the third and fourth spheres, Eudoxus could have produced, in an approximate fashion, the observed retrograde motions. The diagram presented here, "Retrograde Motion from Concentric Spheres," is only a crude and imaginative suggestion of what might have been Eudoxus's system. The actual details of his system have been lost to time.

Retrograde motion can be produced from combinations of spheres rotating with constant velocities. Even four concentric spheres, however, cannot simultaneously produce with quantitative accuracy both the length of the retrograde motion westward and the length of the motion in latitude (north–south) for all the planets.

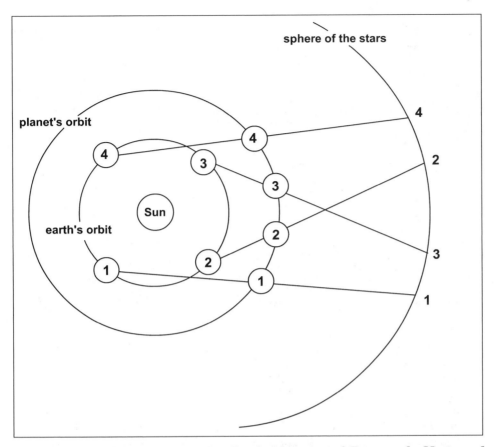

Figure 4.2: Modern Explanation for the Appearance of Retrograde Motion of a Planet. As seen from the Earth at times 1, 2, 3, and 4, the planet apparently moves against the sphere of the stars from 1 to 2, turns back to 3, and then resumes its forward motion to 4. The Earth, moving faster than the outer planet, overtakes and passes it.

Evidently Eudoxus's contemporaries detected flaws in his system, because there occurred a series of modifications after his death, modifications constituting what Thomas Kuhn might characterize as *normal science*. The first modification was made at Eudoxus's school in Cyzicus by his pupil Polemarchus. A second modification was made by Polemarchus's pupil Callippus. They continued their efforts after moving to Athens, where Callippus also worked with Aristotle. In his *Metaphysics*, Aristotle wrote: "Callippus made the position of the spheres the same as did Eudoxus and assigned the same number as did Eudoxus to Jupiter and to Saturn; but he held that two more spheres are to be added to the Sun as well as to the Moon, if one is to account for the phenomena, and one more to each of the other planets" (*Metaphysics*, I8: 1073b17–1074a15).

A system of four concentric spheres can, in principle, give a satisfactory account of the actual motions (in longitude, in latitude, and retrograde) of Jupiter and Saturn, and of Mercury to some extent. For Venus, however, and even more so for Mars, combinations of four concentric spheres produce larger

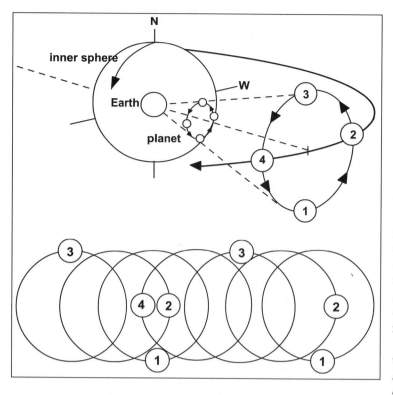

Figure 4.3: Retrograde Motion from Concentric Spheres. The inner sphere is centered on the Earth. Its axis of rotation is horizontal and in the plane of the diagram. As the inner sphere rotates it carries the planet up and down and into and out of the plane of the diagram.

The outer sphere is not visible in the diagram. It, too, is centered on the Earth; the inner and outer spheres are concentric. The outer sphere's axis of rotation is vertical and in the plane of the diagram. As it rotates the outer sphere carries everything within it from right to left.

As the inner sphere moves the planet from 2 to 3 to 4 and the outer sphere moves everything also from right to left, the combined motion of the planet is rapid. But if the speed of the left to right motion from 4 to 1 to 2 imparted by the inner sphere is greater than the steady right to left speed imparted by the outer sphere, then the planet will appear to slow down and briefly move to the right during the passage from 4 to 1 to 2. *Illustration after V. A. Mann, in Norriss S. Hetherington, Astronomy and Civilization (Tucson: Pachart Publishing House, 1987), p. 125.*

deviations in latitude than are actually observed. Callippus was satisfied with Eudoxus's system for Jupiter and Saturn, but he added an additional sphere for each of the other planets.

Neither Aristotle nor Simplicius provides much detail about Eudoxus's system. They report planetary periods only in round numbers of years. It may be that Eudoxus's theory was not fully determined quantitatively. He might have been content solving the problem of retrograde motion qualitatively and geometrically. Determining the various curves traced by a point on the innermost sphere could have been so stupendous and absorbing that it provided its own justification. Lack of descriptive and predictive accuracy would not necessarily have weighed heavily against a theory intended primarily to give conceptual unity to the celestial motions (i.e., to show in a general way that a small number of principles could account for a large number of phenomena). It is tempting, however, to combine with ancient passages a modern analysis of the potential of a system of concentric spheres, then assume that ancient astronomers were in possession of accurate observational parameters, and conclude, with much more certainty than is warranted, that we now understand what was taking place in the minds of astronomers more than two thousand years ago.

Aristotle adopted the ingenious and beautiful geometrical scheme of Eudoxus and Callippus, but he also thought of the spheres as material bodies. The problem for Aristotle was to connect all the spheres physically, yet prevent or compensate for transmission of the motions of outer planets to inner planets. To accomplish this objective he added more spheres to the system. In his *Metaphysics*, Aristotle wrote: "However, if all the spheres combined are to account for the phenomena, there must be for each of the planets other spheres . . . moving counter to these and bringing back to the same position the outermost sphere of the star [planet] . . . ; for thus alone can all the movements combine to produce the complex movement of the planets" (*Metaphysics*, l8: 1073b–1074a15).

Saturn was the outermost planet then known. It was acceptable in planetary models to transmit Saturn's daily rotation around the heavens due to its outermost sphere to Jupiter, but not to transmit inward the motions of Saturn's other three spheres. Aristotle inserted between Saturn and Jupiter three spheres, each rotating around the pole of one of Saturn's inner three spheres, with an equal and opposite velocity. Each of these added spheres neutralized the motion of its corresponding sphere among Saturn's. Thus none of Saturn's motion (except that of its outer sphere) was transmitted down to Jupiter's system of spheres.

Similarly, Aristotle added neutralizing spheres between each subsequent set of planetary spheres and also for the Sun. But none were added for the Moon, because there were no planetary spheres below it to protect from transmission of forces from above. To Callippus's 33 spheres (4 each for Saturn and Jupiter; 5 each for Mars, Mercury, Venus, the Sun, and the Moon) Aristotle added 22 neutralizing spheres (3 each for Saturn and Jupiter; 4 each for Mars, Mercury, Venus, and the Sun). This made a total of 55 (or 56, counting the outermost sphere of the stars), and Aristotle wrote that the total number of moving and countermoving spheres was 55.

So far so good. But Aristotle seemingly continued: "But if one does not add to the Sun and to the Moon the movements we have suggested, all the spheres will number only forty-seven. So much for the number of the spheres" (*Metaphysics*, l8: 1073b17–1074a15). Callippus and Aristotle had added 2 spheres each for the Sun and the Moon to Eudoxus's 3 each, which further required 2 neutralizing spheres for the Sun (but none for the Moon). Thus the total added for the Sun and the Moon was 6 spheres, not the 8 implied by reduction from 55 to 47. Might Aristotle have overlooked the fact that the Moon's motions need not be neutralized and that he had not added two neutralizing spheres for the Moon? Or might some scholar at a later date, not fully understanding what he was writing about and overlooking the fact that the Moon's motions need not be neutralized, have "corrected" what he mistakenly thought was an error by Aristotle. Or might a mistake in the long chain of copying and translation somewhere have changed 49 to 47? Or could the 47 be correct and our understanding in some way deficient?

Eudoxus's system was improved and brought into better agreement with observed planetary motions. There was one phenomenon, however, for which it could not account. The planets move at different times closer to and farther from the Earth. In the case of the Moon, apparent changes in size (due to changing distance from observers on the Earth) were observed directly by the Greeks. For the planets, apparent changes in size, and hence in actual distance, were inferred from changes in apparent brightness. (According to Aristotle, there was no change in the heavens; thus an apparent change in brightness was not real and was explained by a change in distance from the observer.) A system of concentric spheres cannot produce changes in distances of objects from the center of the spheres, where the Earth was assumed to reside.

Simplicius wrote in his commentary on Aristotle's *On the Heavens:*

Nevertheless the theories of Eudoxus and his followers fail to save the phenomena, and not only those which were first noticed at a later date, but even those which were before known and actually accepted by the authors themselves. . . . I refer to the fact that the planets appear at times to be near to us and at times to have receded. . . . The Moon also, even in the perception of our eye, is clearly not always at the same distance from us. . . . The same fact is moreover confirmed if we observe the Moon by means of an instrument; for it is at one time a disc of eleven fingerbreadths, and again at another time a disc of twelve fingerbreadths. . . . Polemarchus of Cyzicus appears to be aware of it [this inequality in the distances of each star (planet) at different times] but to minimize it as being imperceptible, because he preferred the theory which placed the spheres themselves about the very center in the universe. Aristotle, too, shows that he is conscious of it when, in the Physical Problems, he discusses objections to the hypotheses of astronomers arising from the fact that even the sizes of the planets do not appear to be the same always. In this respect Aristotle was not altogether satisfied with the revolving spheres, although the supposition that, being concentric with the universe, they move about its center attracted him. (Heath, *Aristarchus of Samos,* 221–23)

Greek astronomers after Eudoxus accepted the issue of planetary distances as a legitimate problem. Concentric spheres could not account for observed changes in distances, and eventually they were abandoned.

Nonetheless, Eudoxus's planetary model and its continuation by Callippus are impressive. Eudoxus went beyond mere philosophical speculation about the construction of the universe and attempted to account for planetary motions with a geometrical model. Callippus supplied observational facts necessary to test the theory, and he modified it, bringing it into better agreement with observation. This early example of continuity in science illustrates the cumulative advance possible when a problem receives continued attention. Astronomy was on its way to becoming an exact science and one of the most impressive achievements and legacies of Greek civilization.

5

ECCENTRICS AND EPICYCLES

Plato initiated the paradigm of uniform circular motion. Working within the paradigm, Eudoxus devised combinations of concentric spheres. The combined motions were intended to mimic observed planetary motions. Combinations of concentric spheres, however, cannot produce changing distances from the Earth. Next, schemes that could do so were developed, though not in Athens.

Not long after Plato and Eudoxus flourished in Athens, the center of scientific activity in the Greek intellectual world shifted to Alexandria. This port city was established by Alexander the Great in 332 B.C. on the western edge of Egypt's Nile River Delta.

When Alexander died in 323 B.C., his generals split his empire into three major kingdoms: Greece, Asia, and Egypt. With its fertile land along the Nile River, Egypt was wealthy and became even more so after the Egyptian ruler Ptolemy I gained possession of Alexander's body and made his tomb at Alexandria a profitable tourist attraction. In another economic coup, Ptolemy I stopped the export of grain until famine abroad brought higher prices.

Ptolemy I used money from these and other enterprises to begin construction of a lighthouse. Nearly 400 feet high when it was completed by Ptolemy II, it was the tallest building on Earth. Even more wondrous, its mirror reflected light (sunlight during the day and fire at night) that could be seen more than 35 miles offshore. The lighthouse was one of the Seven Wonders of the Ancient World—and the only wonder with a practical use. It may have collapsed during an earthquake in A.D. 1303. A French underwater archaeology project begun in 1994 has recovered much material from the harbor, including pieces of the lighthouse.

Ptolemy I also founded the Museum, around 290 B.C. It was home to a hundred scholars subsidized by the government. There were lectures, and specimens of plants and animals were collected for study.

Not to be outdone, Ptolemy II established the Library. Its famous collection of perhaps half a million books was obtained by purchasing private libraries, including possibly Aristotle's. Astronomical instruments were constructed for use at the Library, and the matching of theory with observation was undertaken on a systematic and sustained basis.

In 47 B.C. Julius Caesar arrived at Alexandria, chasing Pompey. Pompey had won fame in battles, including victory over the remnants of Spartacus's army of slaves and the successful siege of Jerusalem. He married Caesar's daughter Julia and joined Caesar in a ruling triumvirate. After Julia and the third member of the triumvirate died, Pompey gained temporary ascendancy in Rome. Caesar, defying orders of the Roman Senate, crossed the Rubicon River from Gaul with his army to battle Pompey. At Alexandria a traitor surprised the fleeing Pompey and delivered his head and signet ring to Caesar's ship.

Caesar tarried in Alexandria. He was a scholar, and the Museum and Library were major attractions for him. Cleopatra smuggled herself into Caesar's presence only after he went ashore.

Caesar selected thousands of books from the Library to take back to Rome, but they were lost in a fire that spread to the docks from the Alexandrian fleet set afire by Caesar. Later, Mark Anthony, who succeeded Caesar after he was assassinated in 44 B.C., may have given Cleopatra over 200,000 scrolls for the Library.

Both the Museum and the Library suffered in the fourth century A.D. Under the Roman emperor Constantine, Christianity triumphed and pagan institutions were destroyed. In A.D. 392 the last fellow of the Museum was murdered by a mob, and the Library was pillaged.

Whatever remained of the Library was further damaged following the Arab conquest of Alexandria in A.D. 640. Three hundred years later, a Christian bishop known for his critiques of Muslim atrocities asserted, without evidence, that the conquering Caliph reasoned that the books in the great library reputed to contain all the knowledge of the world either would contradict the Koran, in which case they were heresy, or would agree with the Koran, in which case they were superfluous. Supposedly it took six months to burn all the books as fuel for the bathhouses of the city.

A new library building rose in Alexandria in 2002, constructed over 12 years at a cost of $210 million by the Egyptian government and UNESCO (the United Nations Educational, Scientific and Cultural Organization). The slanting roof, made of aluminum and glass, looks like a computer microchip. No provision was made for books to fill the building, other than donations. Ironically, modern Alexandria and its new Library are surrounded by widespread illiteracy, Islamic fundamentalism, and cultural repression, including censorship of books.

In ancient Alexandria much prestige was attached to scholarship and scientific research, and the Ptolemies sought thus to enhance their reputations. Plato had lamented that inasmuch as no city held geometry in high regard, inquiries in the subject languished. This deplorable situation was no longer true, at least not in Alexandria.

The problem set by Plato and pursued by Eudoxus and Callippus in Athens, to account for the observed motions of the planets, the Sun, and the Moon with a combination of uniform circular motions, now guided astronomers in Alexandria.

Unfortunately, if all too typically, little of the historical record has survived. Around A.D. 140 Claudius Ptolemy (not related to the rulers of Egypt) summed up previous astronomical work in his *Mathematical Systematic Treatise*. The *Almagest*, as Ptolemy's great work came to be called, was so comprehensive that its predecessors were rendered obsolete; they ceased to be copied and failed to survive.

Of Ptolemy, himself, little is known. He reported observations made between the ninth year of Hadrian's regime (A.D. 125) and the fourth year of Antoninus Pius's reign (A.D. 141) "in the parallel of Alexandria." This could have been at Alexandria, itself, or at Canopus, 15 miles east of Alexandria. One scholar suggests that Ptolemy had his home in Canopus because it offered better possibilities for a quiet life of study than did the noisy capital of the Hellenistic world. Another scholar counters that Canopus was renowned in the ancient world for its dissolute and licentious lifestyle.

One of the very few of Ptolemy's predecessors known by name is Apollonius. He was born in the city of Perga, in what is now Turkey, sometime during the reign of an Egyptian king who ruled from 246 to 221 B.C. Apollonius moved to Alexandria, where he may have studied with pupils of Euclid, famous for his summary of Greek geometry.

Apollonius is famous for his book on conic sections (i.e., parabola, hyperbola, and ellipse: the curves cut from a right circular cone by a plane). The first four parts of his mathematical book have survived in the original Greek, and parts five through seven are preserved in Arabic translations. Part eight is lost, as is everything Apollonius wrote on astronomy.

Ptolemy wrote in his *Almagest* that a preliminary proposition regarding the retrograde motions of the planets was demonstrated by a number of mathematicians, notably Apollonius. Elsewhere in the *Almagest*, without attributing them to any particular individual, Ptolemy described what have become known as the *eccentric* and *epicycle* hypotheses:

[I]t is first necessary to assume in general that the motions of the planets . . . are all regular and circular by nature. . . . That is, the straight lines, conceived as revolving the planets or their circles, cut off in equal times on absolutely all circumferences equal angles at the centers of each, and their apparent irregularities result from the positions and arrangements of the circles . . .

But the cause of this irregular appearance can be accounted for by as many as two primary simple hypotheses. For if their movement is considered with respect to a circle . . . concentric with the cosmos so that our eye is the center, then it is necessary to suppose that they [the planets] make their regular movements either along circles not concentric with the cosmos [eccentric circles], or along concentric circles; not with these [concentric circles] simply, but with other circles carried upon them called epicycles. For according to either hypothesis it will appear possible for the planets seemingly to pass, in equal periods of time, through unequal arcs of the ecliptic circle which is concentric with the cosmos. (*Almagest*, III 3)

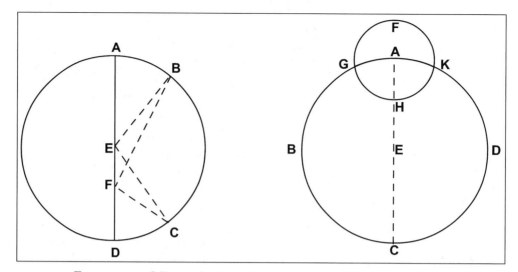

Figure 5.1: Eccentric and Epicycle Hypotheses

In the eccentric hypothesis (above, left), the planet moves around the circle *ABCD* centered on *E* with uniform (unchanging) velocity. The observer, however, is not at the center *E*, but at *F*, from which perspective apparently nonuniform planetary motion is observed.

Ptolemy wrote: "For if, in the case of the hypothesis of eccentricity, we conceive the eccentric circle ABCD on which the planet moves regularly, with E as center and with diameter AED, and the point F on it as your eye so that the point A becomes the apogee [point in the planet's orbit farthest from the Earth/observer at F] and the point D the perigee [point in the planet's orbit closest to the Earth/observer at F]; and if, cutting off equal arcs AB and DC, we join BE, BF, CE, and CF, then it will be evident that the planet moving through each of the arcs AB and CD in an equal period of time will seem to have passed through unequal arcs on the circle described around F as a center. For since angle BEA = angle CED, therefore angle BFA is less than either of them, and angle CFD greater" (*Almagest*, III 3).

In the epicycle hypothesis (above, right), the planet is carried around the small circle *FGHK* with uniform velocity, while that circle is simultaneously carried around on circle *ABCD*, also with uniform velocity. The combined motion of the planet as observed from *E* is not uniform.

Ptolemy wrote: "And if in the hypothesis of the epicycle we conceive the circle ABCD concentric with the ecliptic with center E and diameter AEC, and the epicycle FGHK carried on it on which the planet moves, with its center at A, then it will be immediately evident also that as the epicycle passes regularly along the circle ABCD, from A to B for example, and the planet along the epicycle, the planet will appear indifferently to be at A the center of the epicycle when it is at F or H; but when it is at other points, it will not. But having come to G, for instance, it will seem to have produced a movement greater than the regular movement by the arc AG; and having come to K, likewise less by the arc AK" (*Almagest*, III 3).

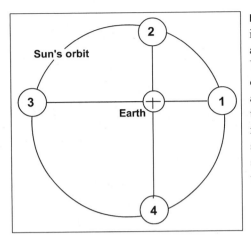

Figure 5.2: Eccentric Solar Orbit. The Sun moves in its orbit with constant speed. It traverses equal distances along the circumference of the circle in equal times. Viewed from the Earth rather than from the center of its circle, the Sun is seen to move through the 90-degree angle from *1* to *2* faster (in less time) than it moves through the 90-degree angle from *2* to *3*. The journey from *3* to *4*, cutting off another 90-degree angle, consumes even more time. Thus the Sun's constant angular speed relative to the center of its orbit appears irregular when viewed from any other reference point.

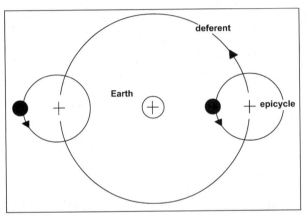

Figure 5.3: Change of Distance in the Epicycle Hypothesis. The large circle (the deferent) is rotating counterclockwise around the Earth at its center. This rotation carries the small circle (the epicycle) attached to the deferent from the left side of the drawing to the right side. Were the epicycle not rotating about its center, the planet carried on the epicycle would end up on the far right side of the drawing as the deferent rotated through an angle of 180 degrees (half a circle) and would always be the same distance from the Earth. If, instead, the epicycle is rotating about its center, through 180 degrees in the same time that the deferent rotates through 180 degrees, the planet will end up *closer to the Earth*, as shown in the above diagram.

Either eccentrics or epicycles, each carrying planets around with uniform circular motions, can produce seemingly irregular motion and thus save the planetary phenomena.

The changeable velocity of the Sun, ignored by Eudoxus, is easily accounted for qualitatively by the eccentric hypothesis. Uniform circular motion (as measured by constant angular velocity about the center of a circle and also as measured by constant velocity along the circumference of the circle) appears nonuniform to an observer not at the center. Thus the Sun appears to an observer not at the center of its circle to move sometimes faster and sometimes slower. Also, the Sun's distance from an eccentric observer changes.

The Sun's distance from an observer on the Earth also changes in the epicycle hypothesis. The large circle (the deferent) rotates around the Earth. Attached to the deferent circle and carried about by it is the center of the epicycle (the small circle). The planet is attached to and carried around by the epicycle rotating around its center. The epicycle thus moves the planet (or the Sun or the Moon) alternately closer to and farther from the Earth.

Gears from the Greeks

In 1900 a sponge diver discovered an ancient shipwreck at a depth of about 140 feet off the tiny Greek island of Antikythera. Along with marble and bronze statues of nude women, jewelry, pottery, and amphorae of wine were a few corroded lumps of bronze. Analysis of the pottery and amphorae suggested that they came from the island of Rhodes and that the ship, probably sailing to Rome with its cargo, sank around 65 B.C., plus or minus 15 years.

X-ray photographs of the corroded lumps of bronze later revealed about 30 separate metal plates and gear wheels. They would have fit into a wooden box about the size of a shoe box. An inscription on one of the metal plates is similar to an astronomical calendar written by someone thought to have lived on Rhodes about 77 B.C. This evidence pretty much ruled out the possibility that the clockwork mechanism had been dropped onto the wreck at a later date, or that it had been left behind by alien astronauts!

Clearly it was some sort of astronomical device, perhaps for navigation. Or possibly it was a small planetarium. The Roman Cicero had written in the first century B.C. about an instrument recently constructed by Poseidonius, which at each revolution reproduced the same motions of the Sun, the Moon, and the five planets. Archimedes was also said to have made a small planetarium, before 200 B.C. Maybe such devices actually existed.

Cleaning away the corrosion of centuries proceeded slowly. Not until the 1950s was a more detailed analysis begun. In the 1970s high-energy gamma rays were used to examine the interiors of the clumps of corroded bronze. Then similarities were revealed between the device, now in the Greek National Archaeological Museum in Athens, and a thirteenth-century-A.D. Islamic geared calendar-computer in the Museum of History of Science at Oxford, showing on dials various cycles of the Sun and the Moon. (The Arabs then had access to ancient Greek texts now lost to us.) Furthermore, both inscriptions and gear ratios from the ancient device were linked to astronomical and calendar ratios.

Still, most historians doubted the conclusion. Certainly the ancient Greeks had possessed the theoretical knowledge necessary to have built such a device, but this was the first physical evidence suggesting that they had attained such an advanced technology. Literary evidence in the form of ancient written accounts of Rhodes' military technology, including a machine gun catapult with gears powering its chain drive and feeding bolts into its firing slot, had been largely ignored or disbelieved.

In 1983 the Science Museum in London acquired a sixth-century-A.D. device that seemingly is one of the missing links between the Greek mechanism of the first century B.C. and the thirteenth-century-A.D. Islamic computer now at Oxford. More recently, in 2002, a new analysis of the Greek mechanism suggests that in addition to accounting for the motions of the Sun and the Moon, it may also have reproduced the motions of the planets using the epicycle model—just as Cicero wrote almost two thousand years ago about a mechanism constructed by Poseidonius.

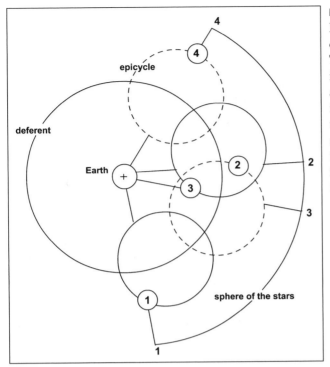

Figure 5.4: Retrograde Motion in the Epicycle Hypothesis. The large deferent circle rotates around the Earth. The center of the small epicycle circle is attached to the deferent and carried around by it. The epicycle rotates around its own center. The planet is attached to the epicycle and carried around by it. As seen from the Earth, the planet appears to move against the sphere of the stars from *1* to *2*, back to *3*, and then resumes its forward motion toward *4*.

Retrograde motion is the appearance of a planet slowing down in its orbit, stopping, briefly reversing course, and then turning to resume its path once more around the heavens. Retrograde motion also can be saved, or reproduced with a combination of regular circular motions, with the epicycle hypothesis, at least qualitatively. A suitable combination of deferent and epicycle sizes and uniform angular velocities can produce the appearance of a planet moving irregularly against the sphere of the stars. In actual practice, quantitatively reproducing the observed widths and spacings of the retrograde arcs is far from simple. This difficulty, however, did not arise in the initial, qualitative period of Greek geometrical astronomy.

6

EQUIVALENCE

Greek geometrical astronomers realized early in the game that their eccentric and epicycle hypotheses produced equal results. The models are equivalent when it comes to reproducing observed phenomena.

Equivalence may appear at first glance a rather esoteric, even frivolous, issue. Indeed, it was dismissed as idle speculation by Theon of Smyrna (now Izmir, Turkey's second-largest port) in his second-century-A.D. handbook of citations from earlier sources on arithmetic, music, and astronomy: mathematical subjects useful for the study of Plato. Theon concluded: "No matter which hypothesis is settled on, the appearances will be saved. For this reason we may dismiss as idle the discussion of the mathematicians [geometrical astronomers], some of whom say that the planets are carried along eccentric circles only, while others claim that they are carried by epicycles, and still others that they move around the same center as the sphere of the fixed stars" (Duhem, *To Save the Phenomena*, 9).

Equivalence held little interest for Theon, but it can illuminate issues of interest to historians and philosophers of science. The mathematical equivalence of two different theories raises the question of which model corresponds to physical reality and, more generally, how to choose between two scientific theories that are observationally equivalent. Furthermore, focusing attention and analysis on instances in which observations do not fully determine theory choice may reveal hidden themata constraining or motivating scientists and affecting their quasi-aesthetic choices.

According to Ptolemy, Apollonius had shown equivalence for a single case. Ptolemy, himself, demonstrated more generally that the identical phenomena will result from either hypothesis. Ptolemy's geometrical demonstration of equivalence occurred in his discussion of solar motion, whose theory he attributed to Hipparchus.

Hipparchus and the Statue of Atlas

No one anticipated a new discovery in the twenty-first century about Hipparchus's long-lost star catalog compiled more than two thousand years earlier, around 127 B.C. Yet in 2005 a pictorial presentation of Hipparchus's data was discovered. It had been in plain sight for centuries for anyone to see, just sitting there, waiting.

The National Archaeological Museum in Naples, Italy, has a seven-foot-tall statue of the god Atlas, who was sentenced by Zeus to hold up the sky. This Atlas is kneeling and holding a globe with 41 constellations on it. The statue is thought to be a late Roman copy of an earlier, now unknown Greek statue (presumably without the fig leaf added by the Romans).

The original Greek sculptor may have copied a globe made by Hipparchus showing positions of stars in his catalog. Historians think that Hipparchus made celestial globes, because ancient coins show him seated in front of a globe and Ptolemy wrote that Hipparchus made such globes.

The constellations chiseled on this marble globe carried by Atlas are placed where they would have been in the sky around 125 b.c. Historians of science can calculate this date because precession, caused by a wobble in the Earth's axis, moves stars around the sky in a 26,000-year cycle. It's like a hand on a clock moving around the sky telling the time. Speculations attributing the positions to Ptolemy's star catalog of about a.d. 150 can now be rejected.

This statue is now recognized as a potential source of ancient data. Further analysis may reveal what Hipparchus used for stellar coordinates and may also help resolve arguments over which stars Hipparchus observed and which Ptolemy observed himself rather than borrowed from Hipparchus.

Little is known about Hipparchus. He may have been born in Nicaea (now northwestern Turkey), because an ancient coin from there depicts him. Ptolemy wrote of 265 years between Hipparchus' and his own observations reported in a star catalog for the year A.D. 137, and also cited astronomical observations made by Hipparchus between 147 and 127 B.C., mainly at Rhodes. Presumably, Hipparchus moved there from Nicaea. Ptolemy also cited a star catalog by Hipparchus, long thought lost.

Theon credited Hipparchus with bringing attention to the equivalence of the eccentric and epicyclic hypotheses:

Hipparchus singled out as deserving the mathematician's attention the fact that one may try to account for phenomena by means of two hypotheses as different as that of eccentric circles and that which uses concentric circles bearing epicycles. The epicyclic appears to be the more common, more generally accepted, and better conformed to the nature of things. . . . For the epicycle is . . . that circle which the planet traces out as it moves on the sphere, . . . the eccentric is altogether different from the circle which conforms to nature, and it is traced out 'accidentally.' Hipparchus, convinced that this is how the phenomena are brought about, adopted the epicycle hypothesis as his own. (Duhem, *To Save the Phenomena*, 8–9)

Ptolemy made the opposite choice. It seemed more probable to him to associate the motion of the Sun with the eccentric hypothesis because it was simpler and was performed by means of one motion instead of two. Simplicity was an important criterion for Ptolemy in evaluating theories. Earlier in the *Almagest*, rejecting Hipparchus's suspicion of a variation in the Sun's motion, Ptolemy had explained: "And in general, we consider it a good principle to explain the phenomena by the simplest hypotheses possible, in so far as there is nothing in the observations to provide a significant objection to such a procedure" (*Almagest*, III 1). Ptolemy explicitly relied on simplicity: one motion rather than two for the Sun.

Ptolemy's choice of eccentric over epicycle seemingly was a straightforward use of the criterion of simplicity. More generally, however, the concept and demarcation of simplicity (and also beauty) often are found to reside in the eye or heart of the beholder. Neither eccentric nor concentric-plus-epicycle is mathematically any simpler than the other.

Ptolemy's choice of eccentric over epicycle, ostensibly based on simplicity, might have been guided by some other consideration, perhaps physical theory. Theon said as much about Hipparchus, namely that he chose the epicycle hypothesis because it better conformed to the nature of things. Astronomers conformed their schemes to Plato's paradigm of uniform circular motions. Were there other beliefs about the nature of things, about essences of bodies and causes of effects, involving aspects of Aristotelian physics, with which astronomical models also had to conform?

7

ASTRONOMY AND
PHYSICS

Astronomers labored to conform their planetary hypotheses to Plato's paradigm of uniform circular motion. Another body of beliefs, more all-encompassing and possibly even stronger than that of uniform circular motion, was Aristotle's physics.

Supposedly, astronomers were not equipped as physicists were to contemplate physical causes and their effects. Nor were astronomers required to derive from the essence of bodies, or from the nature of things, explanations for why other things occurred. In his sixth-century-A.D. commentary on Aristotle's *Physics*, Simplicius wrote about the need for astronomy to conform to geometry and about the difference between astronomy and physics:

> It happens frequently that the astronomer and the physicist take up the same subject
> . . . But in such case they do not proceed in the same way . . . Often the physicist will fasten on the cause and direct his attention to the power that produces the effect he is studying . . . The astronomer is not equipped to contemplate causes . . . He feels obliged to posit certain hypothetical modes of being . . . Whether one assumes that the circles described by the stars are eccentric or that each star is carried along by the revolution of an epicycle, on either supposition the apparent irregularity of their course is saved. The astronomer must therefore maintain that the appearances may be produced by either of these modes of being. (Duhem, *To Save the Phenomena*, 9)

Simplicius concluded:

> This is the reason for Heraclides Ponticus's contention that one can save the apparent irregularity of the motion of the Sun by assuming that the Sun stays fixed and that the Earth moves in a certain way. The knowledge of what is by nature at rest and what properties the things that move have is quite beyond the purview of the astronomer. He posits, hypothetically, that such and such bodies

are immobile, certain others in motion, and then examines with what [additional] suppositions the celestial appearances agree. His principles, namely, that the movements of the planets are regular, uniform, and constant, he receives from the physicist. By means of these principles he then explains the revolutions of all the planets. (Duhem, *To Save the Phenomena*, 9)

Astronomers concerned only with saving the appearances could easily have interchanged the Sun and the Earth in their schemes. According to Simplicius, Heraclides (a student at Plato's Academy and also an attendee at lectures by Aristotle) had done so.

Simplicius could also have mentioned Aristarchus. Again, Ptolemy provides a date: Aristarchus observed the summer solstice at the end of the fiftieth year of the First Kallipic Cycle (279 B.C.). According to Archimedes, who lived in Syracuse (in what is now Sicily) in the third century B.C. and probably studied geometry in Alexandria, "Aristarchus put out a tract of certain hypotheses . . . His hypotheses are that the fixed stars and the Sun remain unmoved, that the Earth revolves about the Sun in the circumference of a circle, the Sun lying in the middle of the orbit. . . ." (Heath, *Aristarchus of Samos*, 302). The tract may have been a book or a written outline or merely explanatory drawings; it may have included some kind of geometric proof or merely stated hypotheses.

Aristarchus has been hailed as an "ancient Copernicus" by antiquarians searching the past for purported promulgations of theories now judged correct. His hypothesis was a remarkable anticipation of the Copernican hypothesis established 1,800 years later. Why, then, wasn't it adopted earlier?

Aristarchus's heliocentric hypothesis claimed no advantage over geocentric hypotheses. And it suffered a considerable potential disadvantage, notwithstanding the distinction being urged by philosophers between astronomy and physics. Any heliocentric hypothesis was incompatible with Aristotelian physics.

The standard interpretation of Aristotle's thought is that he began very close to Plato's intellectual position and only gradually departed from it. Ambiguities in the dating of Aristotle's writings encourage such an analysis, because the resulting pattern can then be used to determine the chronological order of undated passages.

Aristarchus—Heretic?

The Greek philosopher Plutarch (first–second century A.D.) reported that the earlier Greek philosopher Cleanthes (fourth–third century B.C.) had "thought it was the duty of Greeks to indict Aristarchus of Samos on the charge of impiety for putting in motion the Hearth of the Universe, this being the effect of his attempt to save the phenomena by supposing the heaven to remain at rest and the Earth to revolve in an oblique circle, while it rotates, at the same time, about its own axis" (Plutarch, *On the Face in the Orb of the Moon*, VI 922f–923a).

This passage has tempted some historians to imagine a clash between science and religion in the same league as the much later Galileo affair. There is no evidence, however, that anyone ever acted on the purported suggestion to indict Aristarchus for heresy—for disrespecting religious tenets concerning Hestia's Fire or the Earth as a Divine Being. Aristarchus may have been an ancient Copernicus, but he was no ancient Galileo.

A different interpretation of Aristotle's thinking characterizes his philosophy as governed by the interests of a biologist. He continually analyzed and classified, as if what were necessary to understand a subject was to divide it into categories. In his *Poetics*, he divided poetry into tragedy and comedy and analyzed tragedy into six factors. In his *Politics*, he classified the species of government and named their normal and perverted forms (one of which was democracy). In his scientific books, Aristotle took a similar approach.

Aristotle was a member of Plato's Academy for 20 years, yet his philosophy developed very differently. Certainly, he experienced more pleasant encounters with the world of the senses. His statement that the ideal age for marrying is 37 years for the man and 18 for the woman correlates with his own biographical details: at age 37 he married the 18-year-old niece and adopted daughter of the ruler of a small area in Asia Minor.

His father-in-law funded Aristotle's new academy. A few years later, in 342 B.C., Aristotle returned to Macedon, just north of Greece. There he tutored a young prince, Alexander, whose father, Philip, completed his conquest of Greece in 338 B.C. In 336 B.C. Philip was assassinated. Alexander now ruled Greece, and soon much of the Near East. Alexander the Great was more appreciative of Aristotle's tutoring than had been the young Dionysius II of Plato's teaching, and from the Near East sent back a flood of new plants and animals, and also Babylonian eclipse records, to Aristotle in Athens.

A major strength of Aristotle's physical worldview was its completeness. Every part followed logically from the other parts. To understand Aristotle's astronomy, it is necessary to understand his physics. His definitions of motion and space, his conception of what constitutes a cause, and his criterion for an acceptable answer to the question, why? are all essential to an understanding of his view of the physical universe.

In Aristotelian physics, motion was not solely change of position, or *locomotion*, as it was called by Aristotle. He defined motion more broadly as the fulfillment of potentiality. It involved the concept of purposeful action. Aristotle's science is *animistic:* he places the cause of motion within an object, rather than explain motion in terms of outside forces.

In his search for causes, Aristotle often seems content merely to restate effects as undefined powers producing the effects. In Molière's satirical play *Le Malade Imaginaire*, a quack doctor slavishly and stupidly following Aristotelian philosophy explains the sleep-inducing power of opium as due to *virtus dormitiva*, its dormative potency.

Aristotle's sense of motion led him to a particular understanding of place. Place encompassed both motion and potential. Each of the four elements (earth, water, air, and fire) had its natural place. Moved away from its natural place, each element had a natural tendency to return to its natural place.

To explain motion, Aristotle attributed it to a final cause or purpose. An object moved because it had a tendency to return to its natural place, its proper place in the universe. Fire moves upward toward its natural place, and earth falls downward toward its natural place.

Natural motion of a body is movement toward that body's natural place in the universe without the interposition of any force other than the natural tendency to move toward the natural place. The opposite of natural motion was forced or violent motion. It was the result of continuous contact between the moved and a mover.

In his *Physics,* Aristotle discussed the one form of locomotion that could be continuous. Locomotion was either rotatory, rectilinear, or a combination of both. Only rotatory motion could be continuous. Furthermore, rotation was the primary locomotion because it was more simple and complete than rectilinear motion. This value judgment regarding the relative merits of circular and rectilinear motion would affect the subsequent development of planetary astronomy.

Aristotle's views on the organization and structure of the universe are found in his book *De caelo* (*On the Heavens*). It may have been written before his *Physics,* but whatever the chronological order, the logical order is physics followed by astronomy.

Aristotle repeated his belief that locomotion was either straight, circular, or a combination of the two. Bodies were either simple (composed of a single element) or compound. The element of fire and bodies composed of fire had a natural movement upward. Bodies composed of earth had a natural movement downward, toward the center of the universe. Hence, the Earth must be at the center of the Aristotelian universe.

Circular movement was natural to some substance other than the four elements (earth, fire, air, and water). This fifth element was more divine than the other four, because circular motion is prior to straight movement. The fifth element constituted a region beyond the region of the Earth.

Thomas Kuhn has characterized Aristotle's conceptual scheme as a "two-sphere universe" (Kuhn, *Copernican Revolution,* 28). There was a huge sphere of the stars and a tiny sphere of the Earth. The region of change had the Earth in its center, surrounded by water, air, and fire. This region extended up to the Moon. Beyond were the heavenly bodies, in circular motion, with different laws of physics. It was of a superior glory to our region.

And the celestial region was without change. Aristotle wrote: "It is equally reasonable to assume that this body will be ungenerated and indestructible and exempt from increase and alteration . . . The reasons why the primary body is eternal and not subject to increase or diminution, but unchanging and unalterable and unmodified, will be clear from what has been said to any one who believes in our assumptions" (*De caelo,* I3, 269b23–270b16). Aristotle's reasoning may not appeal to modern readers, but his conclusions correctly follow logically from his assumptions. He explained: "If then this body can have no contrary, because there can be no contrary motion to the circular, nature seems justly to have exempted from contraries the body which was to be ungenerated and indestructible, for it is on contraries that generation and decay depend" (*De caelo,* I3, 269b23–270b16).

Comets and new stars, when observed many centuries later, would be placed below the Moon, and any evidence that they were beyond the Moon would pose a challenge to the entire Aristotelian worldview.

Aristotle proceeded to argue that the heavens rotated and that the Earth was stationary in the center. The shape of the heavens was necessarily spherical, because that was the shape most appropriate to its substance and also by nature primary. The heavens also had a smooth finish. The circular motions of the heavens were regular: "Circular movement, having no beginning or limit or middle . . . has neither whence nor whither nor middle; for in time it is eternal, and in length it returns upon itself without a break. If then its movement has no maximum, it can have no irregularity, since irregularity is produced by retardation and acceleration" (*De caelo*, II6, 288a14–289a10).

Aristotle had sought to combine physical dynamics explaining the causes of planetary motions with mathematical kinematics describing planetary motions. This goal proved too ambitious. Neither Eudoxus nor Callippus nor Aristotle succeeded in developing a satisfactory model of concentric spheres. Saving the phenomena mathematically was difficult enough, without complicating the task further with an insistence on a plausible physical mechanism. Greek geometric astronomers generally ignored the direction suggested by Aristotle. Instead, they looked to ingenious and beautiful geometric schemes to save the phenomena, without any confusing physical mecha-

Demonstrative Syllogisms: The Logical Foundation of Aristotle's Science

Aristotle's science was not a science of discovery. It was, rather, a science of demonstration. It was much like a Platonic dialog, the conclusion present from the beginning, if not articulated initially. The goal was to present arguments or phrase questions forcing readers to accept or admit the preordained conclusion.

Aristotle's method of proof was the *syllogism*. He explained that a deduction is a discourse in which, certain things being stated, something other than what is stated follows of necessity from their being so. The following is an example of a syllogism:

All organisms are mortal;
All men are organisms;
Therefore all men are mortal.

The first two propositions are premises. From them necessarily (logically) follows the third proposition, the conclusion. The form is all *A* are *B*, and all *C* are *A*; therefore all *C* are *B*. Demonstrative syllogisms derive facts already known, not new facts. Aristotle established the use of syllogisms in logical presentations.

The best of Greek astronomy was geometric, almost inevitably; geometry was one of the finest intellectual achievements of the Greeks. On the other hand, the Greeks never understood, and in fact distrusted, observation and experiment. Only mathematics, Ptolemy wrote, and he could have included Aristotle's syllogisms, can provide sure and unshakable knowledge. But syllogisms cannot provide new knowledge.

nisms. Still, Greek astronomers were constrained, or guided, by Aristotle's physical theory, especially when choosing between observationally equivalent models. The criterion of which geometric model better conformed to the nature of things was often present, implicitly if not explicitly.

8

SAVING THE PHENOMENA QUANTITATIVELY

Modern scientists try to quantify scientific theories and models, seemingly automatically, without thinking about why they are quantifying or even if they should quantify. A folklorist or an anthropologist looking for and identifying recurring general themes in the preoccupations of scientists might liken their quantitative disposition to a knee-jerk reaction or to the salivating of Pavlov's psychologically conditioned dogs when a bell rang. But if every time they talk or write scientists don't praise explicitly the tremendous power of quantitative reasoning in science, and in many other fields of human thought as well, still they are fully conscious of what they are doing and why. In Greek geometrical astronomy we can find the birth and trace the early development of the quantitative scientific method.

There is no evidence that Apollonius tried to quantify the new eccentric and epicycle models for saving the phenomena. For all we know, he may have been perfectly content with his brilliant qualitative explanations. He may not even have imagined that his geometrical model could take on quantitative, predictive power. Four centuries separate Apollonius's qualitative eccentrics and epicycles from Ptolemy's comprehensive quantitative model of planetary motions. Hipparchus bridges this chasm.

Ptolemy credits Hipparchus with developing the quantitative solar model presented in the *Almagest*, "but under conditions which forced him [Hipparchus], as far as concerns the effect over a long period, to conjecture rather than to predict, since he had found very few observations of fixed stars before his own time" (*Almagest*, VII 1). Ptolemy had more observations, and he corrected Hipparchus's theory. Uncertainty exists over what Ptolemy contributed and what Hipparchus contributed, and some historians attribute to Hipparchus everything not explicitly claimed by Ptolemy as his own.

The primary objective of any solar theory was to demonstrate the motion of the Sun using only uniform circular motions. Ptolemy wrote: "It is now

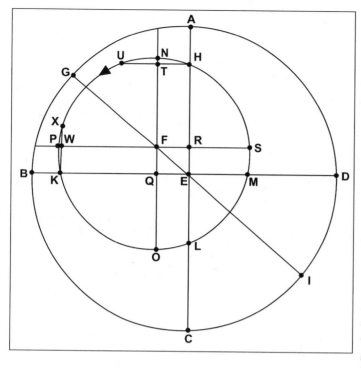

Figure 8.1: Ptolemy's Geometrical Demonstration of the Solar Eccentricity and Apogee. The ecliptic (for Ptolemy, the apparent path of the Sun among the stars; in modern theory, the plane of the Earth's orbit around the Sun marked against the stars) is *ABCD*. It is centered on the Earth at *E*.

The diameters of the ecliptic circle *AC* and *BD* are perpendicular to each other and pass through the tropic and equinoctial points. Spring equinox (one of two points where the ecliptic and the plane of the Earth's equator intersect, as marked on the sphere of the stars; about March 21) is at *A*. Summer solstice (one of two points on the ecliptic where the Sun is at its greatest distance above or below the plane of the Earth's equator; about June 22 in the northern hemisphere, the beginning of summer, and the longest day of the year) is at *B*. Autumnal equinox (about September 22) is at *C*. Winter solstice (about December 22; the shortest day of the year) is at *D*.

If the Sun traveled with uniform motion (constant velocity) around the ecliptic circle *ABCD*, as observed from its center *E*, the Sun would traverse arcs *AB*, *BC*, *CD*, and *DA* in equal times. The seasons (spring, summer, fall, and winter) would be equal in length, a quarter of a year each.

From observations, however, as observed from *E*, arc *HK* is $94\frac{1}{2}$ days; *KL* is $92\frac{1}{2}$; *LM* is $88\frac{1}{8}$; and *MH* is $90\frac{1}{8}$, with *HKLM* an eccentric circle centered on *F*. Uniform motion around *HKLM* appears nonuniform to an observer at *E*.

To make arc *HK* the longest of arcs *HK*, *KL*, *LM*, and *MH*, as seen from *E*, it is readily obvious that the center *F* of the eccentric circle *HKLM* must be placed somewhere in quadrant *AB*. Ptolemy worked out geometrically and quantitatively precisely where in quadrant *AB*.

necessary to take up the apparent irregularity or anomaly of the Sun . . . And this can be accomplished by either hypothesis: (1) by that of the epicycle . . . But (2) it would be more reasonable to stick to the hypothesis of eccentricity which is simpler and completely effected by one and not two movements" (*Almagest*, III 4).

Ptolemy further demanded that theory determine the precise position of the Sun for any given time: "The first question is that of finding the ratio of eccentricity of the Sun's circle . . . what ratio the line between the eccentric circle's center and the ecliptic's center . . . has to the radius of the eccentric circle; and next at what section of the ecliptic the apogee [the point farthest from the Earth] of the eccentric circle is to be found" (*Almagest*, III 4).

This Hipparchus had done: "For having supposed the time from the spring equinox to the summer tropic [solstice] to be $94\frac{1}{2}$ days, and the time from the summer tropic to the autumn equinox to be $92\frac{1}{2}$ days, he proves from these appearances alone that the straight line between the aforesaid centers is very nearly $\frac{1}{24}$ of the radius of the eccentric circle; and that its apogee proceeds the summer tropic by very nearly $24\frac{1}{2}°$ of the ecliptic's 360°" (*Almagest*, III 4). Ptolemy's own numerical solutions, using both Hipparchus's data and his own new observations, agreed with Hipparchus's determinations.

For the Moon's motion, Ptolemy used observations available to Hipparchus, Hipparchus's mathematical procedures, and also new observations, made "very accurately" by himself. Ptolemy corrected and improved the lunar theory handed down from Hipparchus.

Next, Ptolemy tackled the planets:

Now, since our problem is to demonstrate, in the case of the five planets as in the case of the Sun and Moon, all their apparent irregularities so produced by means of regular and circular motions (for these are proper to the nature of divine things which are strangers to disparities and disorders) the successful accomplishment of this aim so truly belonging to mathematical theory in philosophy is to be considered a great thing, very difficult and as yet unattained in a reasonable way by anyone. (*Almagest*, IX 2)

Hipparchus, who had demonstrated the hypotheses of the Sun and the Moon, had not, according to Ptolemy, succeeded with the planets:

I consider Hipparchus to have been most zealous after the truth, . . . especially because of his having left us more examples of accurate observations than he ever got from his predecessors. He sought out the hypotheses of the Sun and Moon, and demonstrated as far as possible and by every available means that they were accomplished through uniform circular movements, but he did not attempt to give the principle of the hypotheses of the five planets, as far as we can tell from those memoirs of his which have come down to us, but only arranged the observations in a more useful way and showed the appearance to be inconsistent with the hypotheses of the mathematicians of his time. (*Almagest*, IX 2)

Hipparchus had found the problem of the planets too complex:

For not only did he think it necessary as it seemed to declare that . . . the regressions of each [planet] are unique and of such and such a magnitude . . . but he also thought that these movements could not be effected either by eccentric circles, or by circles concentric with the ecliptic but bearing epicycles, or even by both together. (*Almagest*, IX 2)

Yet Hipparchus was not content to settle for less than a full quantitative solution, nor was Ptolemy:

Hipparchus reasoned that no one who has progressed through the whole of mathematics to such a point of accuracy and zeal for truth would be content to

Geometrical Astronomical Exercises

Given uniform circular motion and an eccentric solar orbit, determine the ratio of the distance between the centers of the eccentric and ecliptic circles and the radius of the eccentric circle, and where the Sun's apogee is, for the following situations:

1. Spring, summer, fall, and winter all are $91^5/_{16}$ days long.
2. Spring and summer are $182^5/_8$ days long, and fall and winter are 0 days long.
3. Spring and summer are $136^{31}/_{32}$ days long, and fall and winter are $45^{21}/_{32}$ days long.
4. Spring and fall are $91^5/_{16}$ days long, summer is $182^5/_8$ days long, and winter is 0 days long (no Christmas here!).

stop at this like the rest; but that anyone who was to persuade himself and those in touch with him would have to demonstrate the magnitude and periods of each of the anomalies [two anomalies in each planet's position: one relative to the Sun's position, the second relative to the ecliptic] by clear and consistent appearances; and, putting both together, he would have to find out the position and order of the circles by which these anomalies are produced and the mode of their movement and finally show about all the appearances to be consistent with the peculiar property of this hypothesis of the circles. (*Almagest,* IX 2)

New here in Greek geometrical astronomy is insistence on a quantitative science. Astronomers must demonstrate quantitatively the size and the period of each planet's position relative to the Sun's position and in the ecliptic. And they must also demonstrate the size and period of each planet's retrograde arc, by means of well-attested observations.

The quantitative attitude may not command as dramatic an appearance as have scientific discoveries and theories throughout history. Nonetheless, the quantitative attitude is an indispensable foundation of modern science and of Western civilization. The transformation from qualitative to quantitative occurred in the study of planetary motions.

9

PTOLEMY'S EXPOSITION OF MATHEMATICAL ASTRONOMY

Hundreds of years of Greek geometrical astronomy was systematized and quantified with rigorous geometrical demonstrations and proofs by Claudius Ptolemaeus around A.D. 140. He did for astronomy what Euclid had done for geometry and earned a reputation as the greatest astronomer of the ancient world.

Ptolemy's mathematical systematic treatise of astronomy, *The Mathematical Syntaxis,* soon attracted the appellation *megiste,* Greek for "greatest." This was transliterated into Arabic and preceded by *al,* Arabic for "the." Ptolemy's complete exposition of mathematical astronomy became, upon passing from Arabic into medieval Latin in A.D. 1175, the *Almagest.* This Latin translation became, in A.D. 1515, the first printed version of the *Almagest.*

The *Almagest* proceeds in logical order, beginning with a brief introduction to the nature of astronomy and a presentation of the necessary trigonometric theory and spherical astronomy. Then come theories of the Sun and the Moon, an account of eclipses (requiring knowledge of the Sun and the Moon), and discussion of the fixed stars (some of their positions determined with respect to the Moon). The final sections treat the planets, observations of them made largely with respect to the fixed stars.

Ptolemy's motivation is revealed in an epigram appearing in some manuscripts of the *Almagest.* Attribution to Ptolemy is plausible, but not certain. The epigram reads: "I know that I am mortal and the creature of a day; but when I search out the massed wheeling circles of the stars, my feet no longer touch the Earth, but, side by side with Zeus himself, I take my fill of ambrosia, the food of the gods" (Gingerich, *The Eye of Heaven,* 4).

At the beginning of the *Almagest,* Ptolemy distinguished between mathematics and physics, and also between them and theology. He would cultivate mathematics, particularly with respect to divine and heavenly things: "Those who have been true philosophers . . . have very wisely separated the theoretical part of philosophy from the practical. . . . We accordingly thought . . . to train our actions . . . upon the consideration of their [whatever things we happen upon] beautiful and well-ordered disposition, and to indulge in meditation mostly for the exposition of many beautiful theorems and especially of those specifically called mathematical" (*Almagest,* I 1).

Physics dealt with the changeable, corruptible world below the Moon. Astronomy, called mathematics by the Greeks, dealt with the eternal and ethereal world of the Moon, the Sun, the planets, and the stars. Mathematics was "the kind of science which shows up quality with respect to forms and local motions, seeking figure, number, and magnitude, and also place, time, and similar things . . . It can be conceived both through the senses and without the senses" (*Almagest,* I 1). Motion of the planets from place to place (local motion) was the subject of mathematics.

Only indisputable geometrical proof could provide sure knowledge. According to Ptolemy: "Only the mathematical, if approached enquiringly, would give its practitioners certain and trustworthy knowledge with demonstration both arithmetic and geometric resulting from indisputable procedures" (*Almagest,* I 1).

Ptolemy linked divine and heavenly things with the unchanging and with the discipline of mathematics, in contrast to physics, which dealt with the changing, or corruptible. Thus, as Ptolemy explained: "And especially were we led to cultivate that discipline [mathematics] developed in respect to divine and heavenly things as being the only one concerned with the study of things which are always what they are . . . eternal and impassible" (*Almagest,* I 1). *Corruptible* was linked with straight movements and *incorruptible* with circular movements: straight on Earth and circular in the heavens.

Continuing his introduction to the *Almagest,* Ptolemy echoed Plato's concern with education. He cited above all other topics the potential of astronomy—dealing as it does with the constancy, order, symmetry, and calm associated with the divine—to make its followers lovers of this divine beauty and to reform their nature and spiritual state. Ptolemy wrote: "And indeed this same discipline would more than any other prepare understanding persons with respect to nobleness of actions and character by means of the sameness, good order, due proportion, and simple directness contemplated in divine things, making its followers lovers of that divine beauty, and making habitual in them, and as it were natural, a like condition of the soul" (*Almagest,* I 1). This value judgment regarding unchanging versus changing is repeated in the concluding paragraph of the *Almagest's* preface.

Ptolemy also stated his intention to record everything that had already been discovered and to add his own original contributions: "And so we ourselves try to increase continuously our love of the discipline of things which are always what they are, by learning what has already been discovered in such sciences . . . and also by making a small original contribution . . . we shall only report what was rigorously proved by the ancients, perfecting as far as we can what was not fully proved or not proved as well as possible" (*Almagest,* I 1).

He would begin with reliable observations and then attach to this foundation a structure of ideas to be confirmed using geometrical proofs. Concluding his preface and moving on to the order of the theorems, Ptolemy wrote: "And we shall try and show each of these things using as beginnings and foundations for what we wish to find, the evident and certain appearances from the observations of the ancients and our own, and applying the consequences of these conceptions by means of geometrical demonstrations" (*Almagest*, I 2).

Replacing the geometrical astronomy he had inherited with an inductive, observational science would have been no less than a scientific revolution. Determining the reliability of observations, however, other than from their agreement with the very theory they were to confirm, would prove a major problem for Ptolemy.

Ptolemy presented preliminary astronomical concepts in Book I of the *Almagest,* including the spherical motion of the heavens and the nature of ethereal bodies to move in a circular and uniform fashion. He also stated that the Earth is spherical, in the middle of the heavens, and very small relative to the size of the universe. Ptolemy appealed primarily to observations rather than logically deducing these concepts from physical principles.

Ptolemy also presented necessary mathematical techniques. How to determine the length of a chord between two points on a great circle, along with a table of chord lengths for each half degree of arc, comprised the remainder of Book I. All of Book II was devoted to the application of trigonometric techniques to an oblique sphere, such as the Earth.

In Book III, Ptolemy took up the problem of the Sun's motion. He described previous observations of the length of the year and summarized in a table the results of the Sun's regular movement. The next task, he wrote, was to explain the apparent irregularity of the Sun's motion as a combination of regular circular motions. Both the eccentric and the epicycle hypotheses produced the observed motion.

Book IV of the *Almagest* is concerned with motions of the Moon, which Ptolemy presented in a table. Complex irregularities called for ingenious solutions were the lunar observations to be saved by a combination of regular circular motions.

Ptolemy determined the variation of the Moon in latitude from three ancient eclipses observed in Babylon and from three among those "most carefully observed" by himself in Alexandria. (Eclipse observations are particularly convenient to work with, because at these times the Moon is known to be on the ecliptic and at a longitude 180 degrees from the Sun.)

Ptolemy reproduced the lunar anomaly in a qualitative way simply by inclining the plane of the lunar deferent to the plane of the ecliptic. More quantitatively, the Moon was observed neither to cross the ecliptic at the same longitude after each revolution nor to return to the same latitude in equal intervals of time. It would have done so had its epicycle revolved counterclockwise around the deferent with the same angular speed with which the Moon moved clockwise around its epicycle. Ptolemy slowed the speed of the Moon around

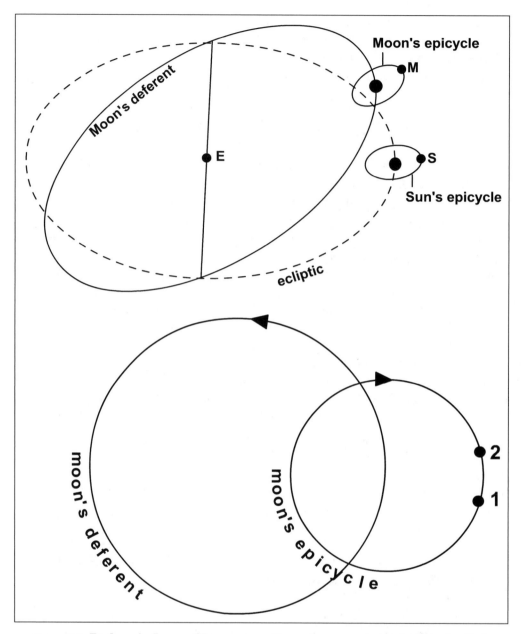

Figure 9.1: Ptolemy's Lunar Theory. top: To produce a variation in latitude, Ptolemy inclined the Moon's deferent at an angle of about 5 degrees to the ecliptic, thus moving the Moon *(M)* above and below the Sun *(S)* as observed from the Earth *(E)*. bottom: A further variation was introduced by reducing the clockwise speed of the Moon around its epicycle. This speed was made less than the counterclockwise speed of rotation of the epicycle itself around the deferent. In the time it took for the epicycle to move through 360 degrees and so return to its original position on the deferent, the Moon, which began at point 1 on its epicycle, moved only to point 2, a clockwise motion of less than 360 degrees.

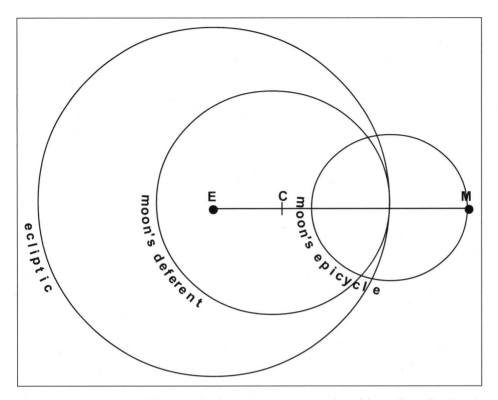

Figure 9.1: (Continued)   above: Yet further variation was produced by making the Moon's deferent, with its center at *C*, eccentric to the Earth, at *E*. This stratagem varied the distance of the Moon's epicycle from the Earth and thus also the apparent angular speed of the Moon's epicycle, *as viewed from the Earth*. However, Ptolemy kept the Moon's deferent rotating with uniform circular motion about the Earth at *E*, rather than about the actual center of the deferent, at *C*.

its epicycle while maintaining the (now greater) speed of the epicycle around the deferent. The resulting combination of regular circular motions adequately reproduced the lunar positions near new and full moon (when the Moon is in the direction of the Sun, and in the opposite direction, as viewed from the Earth).

Agreement was less satisfactory when the Moon was at first or third quarter (90 degrees from the Sun). Ptolemy sought to eliminate this discrepancy by varying the distance of the Moon's epicycle from the Earth (which in turn would vary the apparent angular speed of the epicycle as viewed from the Earth). He placed the Moon's epicycle on a deferent eccentric to the Earth.

The Moon's epicycle now had a new deferent. Ptolemy, however, kept the epicycle's center rotating with uniform circular motion about the center of its previous, geocentric deferent. He did this in order to preserve the match between observation and theory already achieved. The eccentric deferent was also made to rotate with a uniform circular motion about the center of the ecliptic rather than about its own center.

Ptolemy obscured the violation of regular circular motion by using in his example an eccentric circle inside a circle that was itself concentric to the

Historical Significance in Numerical Details

Plodding through mathematical details in the *Almagest* is tiresome for those not imbued with a love of learning for its own sake, but the effort occasionally yields a more tangible reward.

In his discussion of the *synodic* month (the period of revolution of the Moon with respect to the Sun; a *sidereal* month is measured with respect to the stars), Ptolemy noted that there are 4,267 synodic months in 126,007 days and 1 hour. He then concluded that the mean synodic month is therefore 29, 31, 50, 8, 20 days (29 plus 31/60 plus 50/3600, etc., days).

Dividing 126,007 days and 1 hour by 4,267 does not, however, produce this number. See if you don't get, instead: 29, 31, 50, 8, 9.

It so happens that the Babylonian value of the mean synodic month is 29, 31, 50, 8, 20. Obviously, Ptolemy did not do the division implied but instead borrowed the Babylonian value.

Also obvious is that there was an exchange of astronomical information between the Babylonians and the Greeks, an exchange hinting at a more general pattern of mutual influences between Hellenistic and Oriental civilizations.

Dating an Eclipse and its Battle

Reports of ancient observations of the Moon's position can be of interest to historians because these observations hold out hope of dating specific events and, more generally, of relating ancient chronological systems to our own.

We can, for example, calculate past eclipses, particularly large eclipses of the Sun (by the Moon) on a particular day visible at a particular place, and then identify the retrodicted eclipse with a recorded eclipse. This has been done regarding the story of a battle between the Medes and the Lydians in what is now northern Turkey. The battle was brought to an abrupt end by a total solar eclipse, which modern science informs us took place in that region on May 28 in 584 B.C.

ecliptic. In consequence, a straight line from the epicycle's center to the center of the ecliptic cut off equal arcs in equal times on the concentric circle, and a line from the apogee of the eccentric deferent to the ecliptic's center also cut off equal arcs in equal times on the concentric circle. But regular motion of the apogee of the eccentric deferent did not make regular that deferent's revolution about its own center. Furthermore, the regular motion of the epicycle's center with respect to the concentric circle did not translate into regular circular motion with respect to the deferent, the circle on which the epicycle's center actually moved.

Modifications of lunar theory necessary to bring it into agreement with observation violated Ptolemy's own requirement of regular circular motion with respect to the center of the orbit. The violation here, however, was obscure and easily ignored, at least relatively so, in contrast to his discussion later in the *Almagest* of planetary motions.

If all that weren't enough, the motion of the Moon still had to be linked with the motion of the Sun if eclipses were to be accounted for.

After the Moon, Ptolemy moved on to the stars, in Books VII and VIII of the *Almagest*. And after the stars, he moved on to the planets in Books IX through XIII, the last five books of the *Almagest*.

Ptolemy accepted the traditional order of the planets, starting with the Earth in the center of the universe and moving outward: the Moon, Venus, Mercury, the Sun, Mars, Jupiter, and Saturn, all surrounded by the outer sphere of the fixed stars.

Next, Ptolemy restated the basic problem of devising a system of regular circular motions: "Our problem is to demonstrate, in the case of the five planets as in the case of the Sun and Moon, all their apparent irregularities as produced by means of regular and circular motions (for these are proper to the nature of divine things which are strangers to disparities and disorders)" (*Almagest*, IX 2).

The demonstration would be difficult: Ptolemy warned: "The successful accomplishment of this aim as truly belonging to mathematical theory in philosophy is to be considered a great thing, very difficult and as yet unattained in a reasonable way by anyone" (*Almagest*, IX 2). Observations of the planets were available only over short periods, while the error in systematic observations becomes less over longer periods. In addition to uncertain observations, there was more than one anomaly, and the anomalies were intertwined: "In the case of research about the anomalies, the fact that there are two anomalies appearing for each of the planets, and that they are unequal in magnitude and in the times of their returns, works a good deal of confusion. For one of the anomalies is seen to have relation to the Sun, and the other to the parts of the zodiac, but both are mixed together so it is very hard to determine what belongs to each; and most of the old observations were thrown together carelessly and grossly" (*Almagest*, IX 2).

One of the two planetary anomalies (deviations from uniform circular motion) is a planet's seemingly irregular orbital speed, which is linked to its position relative to the ecliptic. The second anomaly is a planet's retrograde motion, which occurs when a superior planet (Mars, Jupiter, or Saturn, in orbit beyond the Earth) is near opposition (on the opposite side of the Earth from the Sun); thus the phenomenon is related to the Sun. The inferior planets (Mercury and Venus, in orbits between the Earth and the Sun) were observed to retrograde when they were in conjunction with the Sun (in approximately the same direction as the Sun, viewed from the Earth).

Note that in the heliocentric model, the inferior planets can never be in opposition. From the Earth, they are always seen near the Sun. In Ptolemy's

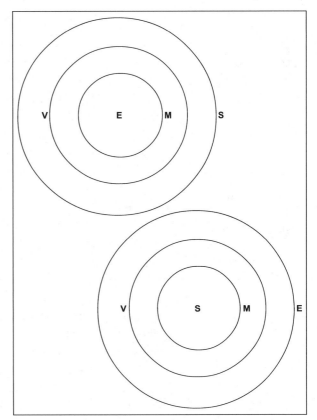

Figure 9.2: Opposition for Inferior Planets. *E* is the Earth; *S*, the Sun; *V*, Venus; and *M*, Mercury.

In a geocentric (Earth-centered) planetary model, an inferior planet (any planet located between the Earth and the Sun) can be in opposition to the Sun (on the opposite side of the Earth from the Sun). In a heliocentric (Sun-centered) model, a minor planet can never be in opposition.

planetary model, however, there was nothing inherent in the geometry to prevent opposition for an inferior planet. He would have to adjust the speeds and starting positions of the inferior planets and the Sun to prevent his model from moving an inferior planet into opposition.

> ## Mental Exercises: Conjunction, Opposition, and Retrograde Motion
>
> Would you expect retrograde motion for a superior planet to occur, if it does occur, when the planet is near opposition or near conjunction? Why?
>
> Would you expect retrograde motion for an inferior planet to occur, if it does occur, when the planet is near opposition or near conjunction? Why?
>
> Can an inferior planet ever be in conjunction? In opposition? Why or why not?

The problem of the planets was difficult, and Ptolemy warned his readers that he might be forced to use something contrary to the general argument, or to presuppose something without immediate foundation, or to devise different types of models for different planets. He argued that "since the appearances relative to the stars are also found to be dissimilar, one should not reasonably think it strange or absurd to vary the mode of the hypotheses of the circles, especially when, along with saving the regular circular movement absolutely everywhere, each of the appearances is demonstrated in its more lawful and general character" (*Almagest,* IX 2).

Planetary observations compiled by Hipparchus revealed that each planet had retrograde arcs that did not occur with constant regularity around the ecliptic, nor were they uniform in size. The epicycle model, however, produces retrograde arcs all the same size and uniformly distributed around the plane containing the deferent and epicycle.

Combining the epicycle and eccentric hypotheses and introducing an eccentric deferent seems, in hindsight, almost inevitably the next step. Although accurate history is not always produced by imagining a logical and possible progression of how science might have developed and then concluding that that is what ancient astronomers did, a plausible case can be sketched.

The eccentric hypothesis was the other type of uniform circular motion that could be called on in an emergency. Furthermore, the retrograde anomalies occurred with respect to the ecliptic, and Hipparchus's solar theory accounted for the anomaly of the Sun's motion, also in the ecliptic, by employing an eccentric deferent circle. And, bottom line, an eccentric deferent does produce nonuniform spacing of the planetary retrograde arcs.

But the size of all the arcs for each planet is still the same. Something more is needed. Maybe even something not in strict accordance with uniform circular motion.

The eccentric hypothesis produces the appearance of greater speed for planets when they are near perigee (the closest point in the planet's orbit to the Earth), and planets consequently traverse their perigee arcs seemingly in less time than the other halves of their orbits. But observations of the planets revealed that the time from greatest speed to mean speed was always greater, not less, than from mean speed to least speed. The epicycle hypothesis, on the other hand, produces the appearance of greater speed at apogee, and correspondingly, the observed phenomenon: that the time from greatest speed to mean speed is always greater than the time from mean speed to least speed.

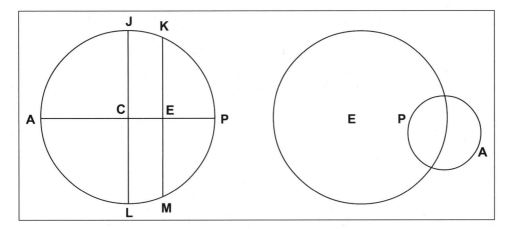

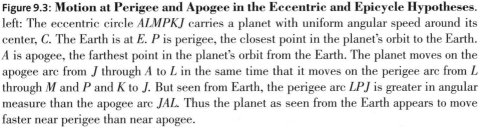

Figure 9.3: **Motion at Perigee and Apogee in the Eccentric and Epicycle Hypotheses**. left: The eccentric circle *ALMPKJ* carries a planet with uniform angular speed around its center, *C*. The Earth is at *E*. *P* is perigee, the closest point in the planet's orbit to the Earth. *A* is apogee, the farthest point in the planet's orbit from the Earth. The planet moves on the apogee arc from *J* through *A* to *L* in the same time that it moves on the perigee arc from *L* through *M* and *P* and *K* to *J*. But seen from Earth, the perigee arc *LPJ* is greater in angular measure than the apogee arc *JAL*. Thus the planet as seen from the Earth appears to move faster near perigee than near apogee.

Mean speed as viewed from the Earth (but understood in this twenty-first-century geometrical construction as the speed that would be viewed from *C*) occurs between *K* and *J* (and also between *L* and *M*) at the point that a line of length *CJ* drawn from *E* would just reach the circumference of the deferent circle between *K* and *J* (and between *L* and *M*). The time from greatest speed to mean speed would be the time consumed moving from *P* to the determined point between *K* and *J*, and the time from mean speed to least speed the time consumed moving from the determined point between *K* and *J* to *A*. The time consumed moving from *P* to anywhere between *K* and *J* will be less than the time consumed moving from anywhere between *K* and *J* to *A*. But, Ptolemy reported that in the case of the five planets the time from greatest speed to mean is always greater than the time from mean speed to least. right: In the epicycle hypothesis, the Earth is at *E*, the center of the deferent circle, which carries around the smaller epicycle, which in turn carries the planet. *A* and *P* are apogee and perigee. The rotations of the deferent and epicycle circles are both counterclockwise. When the planet is near apogee, the two circles' motions are pointed in approximately the same direction, and the motions add together to give a greater resulting speed. When the planet is near perigee, the two circles' motions are pointed in approximately opposite directions, and the motions partially cancel each other, resulting in a lesser net speed. Therefore, in the epicycle model, planets will appear to move faster at apogee and slower at perigee.

There were yet further complications. To deal with them, Ptolemy invented a new concept: the *equant point.* Uniform angular motion, previously defined as cutting off equal angles in equal times at the center of the circle, would now be taken with respect to this new point not at the center of the circle. As Ptolemy earlier had warned, he might be compelled by the nature of his subject to use a procedure not in strict accordance with theory. He explained:

> Now from prolonged application and comparison of observations of individual [planetary] positions with the results computed from the combination of both [the eccentric and epicyclic] hypotheses, we find that it will not work simply to assume . . . that the eccentric circle on which the epicycle center is carried is identical with the eccentric circle with respect to the center of which the epicycle makes its uniform revolution . . . cutting off equal angles in equal times at [that center]. Rather, we find . . . that the epicycle center is carried on an eccentric circle which, though equal in size to the eccentric circle which produces the anomaly, is not described about the same center as the latter. (*Almagest*, IX 5)

The equant point was a questionable modification of uniform circular motion. Planetary epicycles still moved with uniform angular motion but were

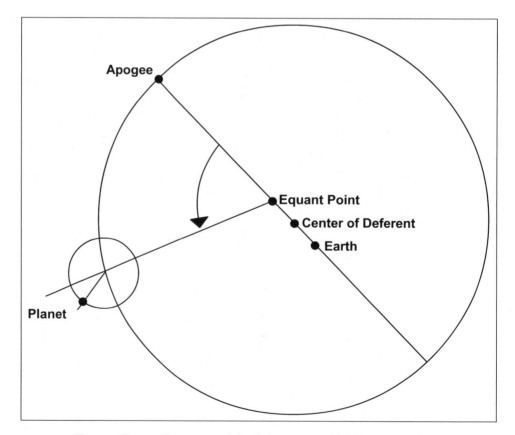

Figure 9.4: Equant Point. The center of the deferent is midway between the Earth and the equant point. The center of the epicycle carrying the planet travels with uniform angular speed as observed not from the center of the deferent but from the equant point.

now measured with respect to points other than the centers of the eccentric circles that carried the epicycles. Ptolemy deemed the equant point necessary to save the phenomena. Fourteen centuries later, Copernicus would condemn the equant point as an unacceptable violation of uniform circular motion.

Ptolemy had set out to write the astronomical equivalent of Euclid's *Elements,* and he succeeded. He showed, for the first time, how to render from observations numerical parameters of planetary models. Unrivaled in antiquity, the *Almagest* was a brilliant synthesis, combining a treatise on theoretical astronomy with practical examples teaching many generations of astronomers how to construct geometric and kinematic models constrained by astronomical observations. The *Almagest* reigned, albeit sometimes in absentia, as the major astronomical textbook, surpassing all that had gone before and not itself surpassed for some fourteen hundred years, until the time of Copernicus.

10

REALITY OR MATHEMATICAL FICTION?

Greek geometrical astronomy is accurately characterized as showing a lack of concern with the physics of the problem of planetary motions and a preoccupation with the mathematics. Given the general acceptance of Aristotle's division between physics and mathematics, and the placement of astronomy within mathematics, it could scarcely have been otherwise.

Some historians and philosophers of science imagine that it was even more. They believe that the models of planetary motions constructed by Greek geometrical astronomers were intended as purely calculating devices having nothing to do with any underlying physical reality. It is easy to agree with this opinion. It is correspondingly difficult to believe in the reality of a universe composed of planets attached to epicycles, themselves rotating uniformly, and also carried about apparently irregularly by eccentric circles actually rotating uniformly around equant points located other than at their centers.

The idea of *instrumentalism* runs through the history of Greek geometrical astronomy. In the instrumentalist view of the relationship between theory and observation, known empirical data are suspended and the study then becomes one of pure geometry, not solving but still relevant to the astronomical problem. All that remains are simple mathematical fictions and pure conceptions, with no question of their being true or in conformity with the nature of things or even probable. For so-called *instrumentalists*, it is enough that a scientific theory yields predictions corresponding to observations. Theories are simply calculating devices.

Realists, on the other hand, insist that theories pass a further test: that they correspond to underlying reality. For realists, Greek geometrical astronomers were describing concrete bodies and movements that actually were accomplished. Realists believe that scientific theories are descriptions of reality. *Dogmatic* realists insist on the truth of a theory. *Critical* realists concede a theory's conjectural character without necessarily becoming instrumentalists.

A disappointed realist may appear to be a *local* instrumentalist with regard to a particular failed theory retaining instrumental value but is far from becoming a *global* instrumentalist.

As early as the sixth century A.D., Simplicius raised the issue of whether the astronomers' combinations of uniform circular motions were real or merely fictions:

> But just as the stops and retrograde motions of the planets are, appearances notwithstanding, not viewed as realities . . . so an explanation which conforms to the facts does not imply that the hypotheses are real and exist. By reasoning about the nature of the heavenly movements, astronomers were able to show that these movements are free from all irregularity, that they are uniform, circular, and always in the same direction. But they have been unable to establish in what sense, exactly, the consequences entailed by these arrangements are merely fictive and not real at all. So they are satisfied to assert that it is possible, by means of circular and uniform movements, always in the same direction, to save the apparent movements of the wandering stars. (Duhem, *To Save the Phenomena*, 23)

Simplicius was considering instances in which the apparent movements were saved. Even stronger candidates for the merely fictive are planetary models that overlooked or ignored known phenomena. Because Eudoxus did not take into account changing distances, some historians and philosophers of science have concluded that his planetary theory was merely a calculating device that had nothing to do with underlying physical realities. It was not true or in conformity with the nature of things or even probable; it was a mathematical fiction. Geometers used it to render celestial motions accessible to their calculations. Hence, Eudoxus was an instrumentalist. Supposedly, scientific theories for him were simply instruments, or calculating devices.

Other historians and philosophers of science argue that Greek geometrical astronomers were describing concrete bodies and movements that actually occurred. Eudoxus's mathematical construction was a theory about the physical world. It was intended to discover and explain the real, rather than the apparent, movements of celestial bodies. Observations functioned as evidential controls on theory; astronomers tried to make the theory fit the observations. Indeed, Eudoxus's system could have been less complex had it been intended merely as a computational device. Hence, Eudoxus was a realist.

Historians and philosophers of science debate whether ancient Greek geometrical astronomers were instrumentalists or realists. A third possibility is that Eudoxus was neither a realist nor an instrumentalist. His system for saving the phenomena was far more than a mere calculating device but less than reality, at least the empirical reality of realists. Eudoxus was infatuated with Plato's paradigm and intellectually imprisoned by it in Plato's cave. He was neither a realist nor an instrumentalist, but a paradigm prisoner.

A modern example of the power of a paradigm or themata is provided by Einstein. He wrote: "I do not by any means find the chief significance of the

Symmetry: A Human and Scientific Value

Another difficulty with equant points, in addition to their sheer physical improbability, was that Mercury's equant point was located differently than were the equant points for all the other planets.

The epicycles were borne with uniform motion on circles the same size as the eccentric circles but with other centers. Ptolemy wrote: "And these centers, in the case of all except Mercury, bisect the straight lines between the centers of the eccentrics . . . and the center of the ecliptic. But in the case of Mercury alone, this other center [equant point] is the same distance from the center revolving it as this center [center of the eccentric] . . . is in turn from the center effecting the anomaly on the side of the apogee. . . ." (*Almagest*, IX 5). Mercury's equant point was located not between the centers of the ecliptic and the eccentric, but beyond the center of the eccentric, in the direction of the apogee, at the same distance from the center of the eccentric as the distance between the centers of the eccentric and the ecliptic. Furthermore, Ptolemy noted: "In the case of Mercury alone, we find the eccentric circle revolved by the aforesaid center, contrariwise to the epicycle, back westward one revolution in a year's time" (*Almagest*, IX 5).

The violation of symmetry between Mercury's motion and the motions of all the other planets would disturb the sensibilities of later astronomers, regardless of the good agreement of Ptolemy's system with observations. Indeed, it was Ptolemy's insistence upon accurately accounting for all the known motions that forced him to compromise on the aesthetics of his system.

Actually, Ptolemy need not have given Mercury a unique mechanism. Often obscured by the Sun's glare, Mercury is the most difficult to observe of the planets visible to the naked eye. Ptolemy was unable to observe Mercury on a crucial occasion because it was too close to the horizon. Assuming a symmetry that does not exist, he erroneously substituted for the unmeasured position another observation made at a different time. And then, trying to accommodate this false data point in his theory, he was led to a unique mechanism for Mercury.

Ptolemy would have been better off had he assumed that all the planets had the same mechanism and then fit his observations of Mercury to such a mechanism as best he could. In this instance, Ptolemy placed too much faith in his own crude observations and too little faith in aesthetic considerations.

general theory of relativity in the fact that it has predicted a few minute observable facts, but rather in the simplicity of its foundation and in its logical consistency" (Hetherington, "Plato's Place," 109). The themata of simplicity was the foundation of Einstein's quasi-aesthetic faith in relativity theory more so than were a few confirming observations. The astronomer Arthur Eddington, purportedly one of only three people who really understood relativity theory, elaborated: "For those who have caught the spirit of the new ideas the observational predictions form only a minor part of the subject. It is claimed for the theory that it leads to an understanding of the world of physics clearer and more penetrating than that previously attained" (Hetherington, "Plato's Place," 109). (On hearing that he was one of only three people to really understand Einstein's relativity theory, Eddington is said to have asked who was the third.)

Perhaps Eudoxus believed his vision of the universe so penetrating that observation was rightly consigned to a minor role. If he did, can modern scientists, given their own heritage in Einstein and Eddington, reject such an attitude as nonscientific? And is it any wonder that Eudoxus defies neat classification as either an instrumentalist or a realist?

Ptolemy, too, has been the subject of attempts to classify him as either an instrumentalist or a realist. Although he compounded many motions to determine the trajectories of planets, he refused to impose contraptions of wood or metal on the motions. His constructions had no physical reality; only the resultant motion was produced in the heavens. Nowhere in the *Almagest*, however, is there any explicit suggestion by Ptolemy that he understood his work as inventing merely mathematical fictions to save the phenomena. Instead he was after reality. Side by side with Zeus himself, so Ptolemy had written, he searched out the massed wheeling circles of the planets.

The equant point may be the final, culminating fiction for readers forced from a realist to an instrumentalist characterization of Ptolemy's work. Yet his need to invent and resort to equant points illustrates his insistence on accurately accounting for all the known planetary motions. The more improbable the equant point is, the more strong the argument that Ptolemy was some kind of realist.

Continuing in the *Almagest* beyond the introduction of equant points, even more material is found to fuel debates over whether Ptolemy was a realist or an instrumentalist. He had begun by treating all the planets as if their orbits were restricted to the ecliptic (the plane of the Earth's orbit around the Sun). There are, however, variations in the latitudes of the planets. Inclining the planets' eccentric circles to the plane of the ecliptic took care of much, but not all, of the observed variations in latitude (north–south, above and below the plane of the ecliptic).

To produce further variation, Ptolemy invented little circles to vary the inclinations of the epicycles to their eccentric circles. He explained that: "In the case of Venus and Mercury . . . those diameters of the epicycles . . . are carried aside by little circles . . . proportionate in size to the latitudinal deviation, perpendicular to the eccentrics' planes, and having their centers on them. These little circles are revolved regularly and in accord with the longitudinal passages" (*Almagest*, XIII 2)

There were yet further complications:

> Now, concerning these little circles by which the oscillations of the epicycles are effected, it is necessary to assume that they are bisected by the planes about which we say the swayings of the obliquities take place; for only in this way can equal latitudinal passages be established on either side of them. Yet they do not have their revolutions with respect to regular movement effected about the proper center, but about another which has the same eccentricity for the little circle as the star's longitudinal eccentricity for the ecliptic. (*Almagest*, XIII 2).

In other words, each little circle had its own equant point.

Realizing that his readers were likely to react unfavorably to these added intricacies, Ptolemy pleaded that when it was not possible to fit the simpler hypotheses to the movements in the heavens, it was proper to try any hypotheses. He wrote: "Let no one, seeing the difficulties of our devices, find troublesome such hypotheses. For it is not proper to apply human things to divine things . . . But it is proper to try and fit as far as possible the simpler hypotheses to the movements in the heavens; and if this does not succeed, then any hypotheses possible" (*Almagest,* XIII 2). Ptolemy continued:

> Once all the appearances are saved by the consequences of the hypotheses, why should it seem strange that such complications can come about in the movements of heavenly things? For there is no impeding nature in them, but one proper to the yielding and giving way to movements according to the nature of each planet, even if they are contrary, so that they can all penetrate and shine through absolutely all the fluid media; and this free action takes place not only about the particular circles, but also about the spheres themselves and the axes of revolution. (*Almagest,* XIII 2)

Had Ptolemy been an instrumentalist interested only in the resulting motion, he would not have had to bother himself about the complicated nature of his devices. Clearly, Ptolemy envisioned actual physical structures in the heavens carrying around the planets, controlling the motions of the planets. The structures were not made of wood or metal or other earthly material, but rather of some divine celestial material offering no obstruction ("there is no impeding nature in them") to the passage of one part of the construction through another.

The argument from planetary latitudes for Ptolemy as a realist is strong but not conclusive, and examination of his lunar theory finds an equally strong argument for him as an instrumentalist. While it predicted the Moon's positions in longitude and latitude accurately, Ptolemy's theory greatly exaggerated the monthly variation in the Moon's distance from the Earth. Therefore—so the argument goes—Ptolemy could not have intended that the theory be interpreted realistically. He had measured the variation in the observed angular diameter of the Moon (a result of variation in distance) and must have known that his theory failed in this aspect. The astronomer and historian of astronomy J.L.E. Dreyer concluded: "It had now become a recognized fact, that the epicycle theory was merely a means of calculating the apparent places of the planets without pretending to represent the true system of the world, and that it certainly fulfilled its object satisfactorily, and, from a mathematical point of view, in a very elegant manner" (Dreyer, *History of the Planetary Systems,* 196).

Unlike Dreyer, Ptolemy is silent on the matter. His silence on this issue could be interpreted as an implicit understanding that his theory was merely a calculating device, but Ptolemy's silence does not prove beyond reasonable doubt that he had set aside the problem of the variation in the Moon's distance as one of purely mathematical complexity. He might instead have silently continued seeking a realistic physical solution, unsuccessfully battling the physical complexity of the situation. A disappointed or temporarily defeated realist is not necessarily an instrumentalist.

Proponents of Ptolemy as a realist cannot entirely deny the argument from lunar distances. They must try, instead, to dilute it or overwhelm it with a mass of contrary evidence. Ptolemy addressed at great length in the concluding section of the *Almagest* the issue of the physical complexity of his system. Furthermore, in one of his other books, the *Planetary Hypotheses,* Ptolemy nested the mechanism of epicycles and deferents for each planet inside a spherical shell between adjoining planets. Even more so than in the *Almagest,* in the *Planetary Hypotheses* Ptolemy revealed his concern with the physical world.

Proponents of Ptolemy as an instrumentalist must resort to a schizophrenic Ptolemy. In the *Planetary Hypotheses,* he is a realist; but earlier, in the *Almagest,* an instrumentalist.

Ptolemy did not push Plato's analogy of the cave to the extreme conclusion that reality exists only in the mind, or that reality is to be found beyond the visible façade of the phenomena in the mathematical structures that generate them. In changing the location of reality from the mind to the heavens, Ptolemy was more of a realist than were Plato and Eudoxus. But if Ptolemy's reality existed in the heavens, it was not a materialistic reality. It was an ethereal reality corresponding to nothing on the Earth.

11

THE GREATEST ASTRONOMER OF ANTIQUITY OR THE GREATEST FRAUD IN THE HISTORY OF SCIENCE?

In the 1970s, Robert R. Newton, a geophysicist at the Johns Hopkins University Applied Physics Laboratory, accused Ptolemy of having committed a crime against his fellow scientists and scholars, and of betraying the ethics and the integrity of his profession. Newton charged that: "Ptolemy is not the greatest astronomer of antiquity, but he is something still more unusual: He is the most successful fraud in the history of science." Furthermore, "the *Almagest* had done more damage to astronomy than any other work ever written, and astronomy would be better off if it had never existed" (Newton, *Crime of Claudius Ptolemy*, 379).

Newton had been attempting to determine the acceleration of the Moon in its orbit around the Earth from a comparison of modern observations against Ptolemy's reported data. (Tidal friction between the seas and the solid Earth slows the Earth's axial spin; and conservation of angular momentum in the Earth-Moon system is conserved as the Moon speeds up.) Because of the role of celestial positions in navigation, the United States Navy was financing the research. Much to his surprise, Newton found that Ptolemy's observations too closely matched his lunar theory. Seemingly, Ptolemy had fabricated his reported observations.

A more recent study calculates how many years must be extrapolated backward in time to obtain Ptolemy's reported lunar positions, assuming that the rate of acceleration of the Moon's motion has remained constant from then to now. By this means, three tables in the *Almagest* are dated, roughly, to 1000 B.C., to 1700 B.C., and to 1800 B.C. These are too old to have been made by Greek astronomers, and may be Babylonian.

Robert Newton might have saved himself much effort and aggravation had he only known a little history. As far back as the tenth century, Arab astronomers had voiced suspicions that Ptolemy might simply have taken over the observations of a predecessor and adjusted them for his own time. Astronomers since then, including the Danish astronomer Tycho Brahe in the sixteenth century, the English astronomer Edmund Halley in the seventeenth century, and the French astronomer Pierre-Simon Laplace in the eighteenth century, all concluded that Ptolemy had borrowed or fabricated his reported observations.

Studies after Newton's analysis of Ptolemy's lunar data also lead to much the same conclusion. For example, regarding the planet Mars, the Harvard astronomer and historian of astronomy Owen Gingerich finds that "although Ptolemy's stated observations are not particularly good . . . they do match his tables in an uncanny way. In fact, this situation prevails throughout the entire *Almagest,* which leads to the suspicion that something is going on that does not meet the eye, something craftily concealed in the writing of the *Almagest*" (Gingerich, *Eye of Heaven,* 17). Also with Mercury, a close examination finds agreement between theory and observation too good to be true.

Ptolemy's star catalog, too, has long been suspect. In addition to small random errors, the catalog has a large, constant systematic error that leaves, on average, the longitudes (east–west positions) too small by about a degree, while the latitudes (north–south positions) are basically correct.

Suppose that Ptolemy believed that precession (a slow change in the orientation of the Earth's axis of rotation) has no effect on latitude but increases the longitudes of stars by 1 degree per century. Might not he then simply have taken over Hipparchus's star catalog of 129 B.C. and added to each of its reported longitudes 2 $^2/_3$ degrees, for the 2 $^2/_3$ centuries between 129 B.C. and A.D. 137, the epoch for which Ptolemy's catalog was prepared?

The actual amount of precession, however, for this time interval is $3^2/_3$ degrees. Hence, Ptolemy's longitudes, if simply Hipparchus's with 2 $^2/_3$ degrees added, would be a degree too small, as in fact they are. Thus the mean error in Ptolemy's longitudes is neatly explained.

The error in longitudes may, however, have a more innocent explanation. Ptolemy might have measured the longitudes of stars relative to a fundamental star, which in turn was measured relative to the Sun (or relative to the Moon, which, in turn, was measured relative to the Sun). The fundamental star's absolute longitude would have been obtained by adding its longitude relative to the Sun and the Sun's absolute longitude as calculated from solar theory. And since Ptolemy's solar theory was incorrect in a way that resulted in too small a solar longitude by about a degree, all his reported stellar longitudes would be off by this same amount.

In the context of precession, Ptolemy has been caught using data selectively. Had he calculated the precession of the equinoxes from all of the 18 easily recognizable stars that he listed in the *Almagest,* Ptolemy would have found approximately the modern value for precession. He reported, however, that one could conclude from the data that precession was 1 degree in about 100 years (exactly what Hipparchus had determined earlier). Next, Ptolemy showed how 6 of the stars supported this conclusion; he did not mention the other 12 stars.

Often in the first enthusiasm of a new discovery, disagreeing facts are simply disregarded. The *Almagest* is Ptolemy's final, formal presentation, however, not the first flash of his enthusiasm.

Another issue in the debate over Ptolemy's star catalog revolves around the fact that not a single star listed was invisible from Rhodes, the home of Hipparchus and 5 degrees north of Alexandria, the purported observational home base for the *Almagest*. An astronomer in Alexandria might (but not necessarily) have observed stars closer to his extended southern horizon. On the other hand, the magnitudes reported by Ptolemy of 6 southern-most stars are consistent, after adjusting for atmospheric extinction, with having been observed in Alexandria in his time. (The observed brightness of stars is decreased by atmospheric extinction, which increases greatly near the horizon.) And 5 of the 6 stars have magnitudes inconsistent with having been observed at Rhodes in Hipparchus's time. (A star near the southern horizon appears much dimmer to an observer at Rhodes than at Alexandria.)

The distribution of fractional values in Ptolemy's stellar longitudes is also a prominent issue in debates over the possible fabrication of observations. In the *Almagest*, Ptolemy reported the latitudes and longitudes of stars as so many degrees plus fractional parts of a degree. The fractions were 0, $\frac{1}{6}$, $\frac{1}{4}$, $\frac{1}{3}$, $\frac{1}{2}$, $\frac{2}{3}$, $\frac{3}{4}$, and $\frac{5}{6}$ degree (corresponding to 0, 10, 15, 20, 30, 40, 45, and 50 minutes).

Given a random distribution, it is equally likely for stars to be reported as located at each division of 5 minutes of the heavens. But with an instrument or instruments calibrated to measure only the above fractions, stars that would otherwise be reported at 5, 25, 35, or 55 minutes cannot be because these "forbidden" slots do not correspond to any of the fractions employed. Instead, the stars must be placed in an adjoining slot. All stars nearest the 5-minute mark must be reported at either 0 or 10 minutes (presumably with a fifty-fifty chance), and all stars at 55 will fall into the 50 or the 0 slot. Therefore, twice as many stars will be reported at 0 as would have been otherwise (all of the stars actually at 0, half of those at 5, and half of those at 55). Similarly, twice as many stars will be reported at 30 minutes, or half a degree, because of the missing 25- and 35-minute coordinates.

Rounding half the stars in the missing 5-minute slot up to the 10 slot gives it half as many stars again as it otherwise would have. The same applies for moving half the stars at 35 to 40. Similarly, half the stars in the missing 25 and 55 slots are rounded down into the 20 and 40 slots. Furthermore, we expect minimums at $\frac{1}{4}$ and $\frac{3}{4}$ degree (15 and 45 minutes, with no rounded up or down additions, because all the adjoining slots are represented by fractional measurements).

Starting with, say, 1,200 stars, or 100 for each 5 minute interval, we would expect 100 each at $\frac{1}{4}$ and $\frac{3}{4}$ degree; 150 each at $\frac{1}{6}$, $\frac{1}{3}$, $\frac{2}{3}$, and $\frac{5}{6}$ degree; and 200 each at 0 and $\frac{1}{2}$ degree. This expectation is realized, in a general way, in Ptolemy's star catalog in the *Almagest* for stellar latitudes. There are only 88 stars at $\frac{1}{4}$ degree, and 50 at $\frac{3}{4}$ degree. At $\frac{1}{6}$, $\frac{1}{3}$, $\frac{2}{3}$, and $\frac{5}{6}$ degree there are 106, 112, 129, and 107 occurrences. There are 236 stars with measurements of 0 fractional degree, and 198 stars with measurements of $\frac{1}{2}$ a fractional degree. This distribution raises no suspicion among historians or astronomers in search of criminal activity.

But look at the distribution of Ptolemy's fractions in stellar longitude. He reports 226 stars at 0; 182 at $\frac{1}{6}$; only 4 at $\frac{1}{4}$; 179 at $\frac{1}{3}$; 88 at $\frac{1}{2}$; 246 at $\frac{2}{3}$; no stars at $\frac{3}{4}$; and 102 stars at $\frac{5}{6}$. How can we explain the peak at $\frac{2}{3}$ degree other than by calling it evidence of a Hipparchian peak at 0, to which Ptolemy, to update the catalog for precession, added $2\frac{2}{3}$ degrees? The other peak, originally at $\frac{1}{2}$ degree, would be shifted to $\frac{1}{6}$ (i.e., $\frac{1}{2} + \frac{2}{3} = 1 + \frac{1}{6}$).

Furthermore, if we increase by $\frac{2}{3}$ degree (40 minutes) the original $\frac{1}{4}$ (15 minute) and $\frac{3}{4}$ (45 minute) slots, they become 55- and 25-minute slots. But these slots do not exist. Their contents might be shifted into neighboring slots. Adding the contents of the 55 slot to the zero slot and the 25 to the 20 slot ($\frac{1}{3}$ degree) produces numbers nearer what Ptolemy reported.

The new $\frac{1}{4}$ and $\frac{3}{4}$ degree categories were produced, we speculate, by shifting forward the original (missing) 35- and 5-minute slots, which had no content, by $\frac{2}{3}$ of a degree (40 minutes). Ptolemy reports only 4 stars with a fractional longitude of $\frac{1}{4}$ degree, and none with $\frac{3}{4}$ degree.

The other original forbidden levels, 25 and 55 minutes, would be shifted to 5 and 35 minutes. These are also forbidden and hence unreported; thus they are not in need of explanation.

Mental Exercise: The Distribution of Measurements

You measure stellar latitudes with a device marked in fifths of a degree, and a colleague measures latitudes of other stars with a device marked in quarters of a degree. Next, you two combine your data.

Assuming a random distribution of the stars, at what fractional degree marking would you expect the peak number of measures to fall?

At what fractional degree marking would you expect the minimum number of measures to fall?

If latitudes for 2,000 stars were recorded, what quantitative distribution of fractional degrees would you expect to see?

What distribution might you expect if the observing devices were marked in fourths and thirds of degrees?

The distribution evidence is suggestive, but not conclusive. Suppose that Ptolemy actually made his reported measurements against a reference star, later found that his longitude for the reference star was short by a fractional amount of $\frac{2}{3}$ degree (or over by $\frac{1}{3}$ degree), and then added this $\frac{2}{3}$ degree to (or subtracted $\frac{1}{3}$ degree from) all his genuine measurements? Or perhaps he only determined the absolute longitude of the reference star after making the relative measurements, and it had a fractional part of $\frac{2}{3}$ degree, which he then added to each star's relative longitude? Indeed, 3 of the 4 bright stars that Ptolemy might conveniently have used as reference stars have fractional $\frac{2}{3}$ degree longitudes in the *Almagest*. The addition of a reference star's fractional longitude can account for the distribution of fractions in longitude, without denying the genuineness of Ptolemy's reported observations.

Interpretations of the distribution of fractional values in Ptolemy's stellar longitudes are numerous and ingenious but so far none has proven decisive. For every conclusion so far proposed, subsequent examination by another historian of astronomy has uncovered one or more possible and plausible alternative explanations.

In too many instances, agreement between Ptolemy's theory and observation is too good to be true. But if there is a hint of fraud in the *Almagest*, there is also scientific greatness. Agreement between Ptolemy's numerical parameters and modern values is too close to be fortuitous. Probably, Ptolemy had a large number of observations, and errors largely canceled each other out in calculations of a general theory. Next, Ptolemy might have selected from among his observations a few in best agreement with the theory, and then presented these examples to illustrate the theory.

Ptolemy lacked our modern understanding of error ranges, standard deviations, and the use of mean values from repeated observations—concepts that would have enabled him to present a theory not necessarily in total and absolute agreement with every data point obtained, but in a less strict agreement with all the data points to within a statistically defined interval around a mean value. Instead, absent any tolerable fluctuation in the agreement between theory and observation, every measurement would be understood by Ptolemy and his contemporaries as an exact result. Consequently, judicious selection from among many measurements was required.

We should also recognize that the *Almagest* is not a modern research paper. Rather, it was a textbook. Ptolemy was attempting to demonstrate a new type of science, in which specific observational data were converted into the numerical parameters of a geometrical model. The lesson taught in the *Almagest* would enable astronomers in the future to add their own observations over a longer temporal baseline and obtain an even more accurate theory or model of planetary motions. Any fudging or fabrication by Ptolemy might well have been understood by him as little fibs allowable in the neatening up of his pedagogy, not as lies intended to mislead his readers about crucial matters. The Greek astronomical tradition was far more concerned with general geometrical procedures than with specific numerical results. Modern norms of science did not yet exist.

Gingerich challenges any attribution of criminal and nefarious motives to Ptolemy but at the same time acknowledges that Newton deserves credit for bringing to our attention inconsistencies and anomalies in Ptolemy's work. History is an activity and an argument, not merely a chronological collection of facts, and in pursuing questions of possible fraud, scholars, including Newton and Gingerich, are making history.

12

ISLAMIC PLANETARY ASTRONOMY

Ptolemy's *Almagest,* the highest achievement of Greek geometrical planetary astronomy, was without rival for some fourteen centuries, until supplanted by Copernicus's *De revolutionibus orbium coelestium* after A.D. 1543. Yet, awe inspiring as the *Almagest* was, and indeed still is, it was far from complete or satisfactory, even by Ptolemy's own standards. Striving valiantly to save the phenomena, whatever the cost in additional circles, Ptolemy had drifted away from a simple planetary model toward an overly elaborate, Rube Goldberg–like monstrosity. (Goldberg, a cartoonist, drew weird contrivances performing simple tasks with many bizarre and unnecessary steps.) Even worse was Ptolemy's introduction of the equant point, a violation of uniform circular motion. Why then was the Copernican revolution, its necessity so evident, so long delayed in coming?

That no man of Ptolemy's genius again walked the Earth for many centuries might suffice as an explanation for the time lag. This sort of explanation is not in current fashion, however. Nor does it illuminate the problem. A hypothesis more open to investigation and more productive of further study is that a whole tradition of astronomy was interred shortly after Ptolemy and not soon resurrected. In the Islamic world, it was not pursued until nearly a millennium later; and in the Latin-reading West not until shortly before Copernicus.

The most obvious cause of disrupted astronomical activity, as well as most other activities of the Roman world, was the decline of that civilization. Also, interest shifted from traditional intellectual pursuits to service in the rapidly growing Christian religion.

This historical explanation presumes existence of an intellectual inertial force and, consequently, continuation of research in geometrical planetary astronomy until disrupted or diverted by an opposing force. We might, however, assume instead that a state of intellectual inattention is the natural state of affairs and then ask whether there were forces present to cause either continuation or revival of work in the astronomical tradition.

Neither historical approach—assuming inertial progress or a natural state of rest—need necessarily be exclusive. Indeed, it is difficult to place unambiguously in a single category all aspects of the long neglect of Ptolemy's *Almagest.*

In the West, the break in the Ptolemaic tradition runs deeper than a general and gradual decline of civilization, a decline so gradual that scientific activity in other fields continued sporadically for centuries. Greek science was distilled into handbooks, and it was primarily through this medium that the late Greek science of Alexandria became known to Latin readers. The *Almagest,* however, was written after the major incorporation of Greek handbook knowledge into Latin. Hence, Ptolemy's achievement was known in the West only by reputation until the twelfth century. Nor did later commentaries on Ptolemy's work by astronomers at Alexandria find their way immediately to the Latin-reading West.

Alexandria already had become a provincial city in the Roman Empire by Ptolemy's time, and in the second and third centuries A.D., political crises and almost incessant civil war in the Empire disrupted internal order and economic production. Among the casualties were the Museum and the Library. The major destruction of these centers of learning occurred in the fourth century when, under Emperor Constantine, Christianity triumphed in the Empire and pagan institutions were destroyed. In A.D. 392 the last fellow of the Museum was murdered by a mob and the Library was pillaged. Whatever may have remained was further damaged in the Arab conquest in the seventh century A.D.

Islam spread rapidly after Mohammed arrived in Medina in A.D. 622, conquering first Mecca, then the rest of the Arabian peninsula—east through what is now Iran, west through North Africa by A.D. 670, and across the Mediterranean to Spain in A.D. 711. This invasion of Europe was only blocked in A.D. 732 at the Battle of Tours. Not until A.D. 1248 were Cordoba and Seville retaken by Christians, and Granada did not fall until A.D. 1492.

Those conquered by Islam were to be left undisturbed in their way of life on condition that they pay tribute. The conquerors' initial intention was not to spread Islam but to substitute one elite for another and to return wealth, in the form of taxes, to the central treasury. Gradually, though, masters and subjects merged. The central administration moved to Damascus and then to Baghdad, vainly hoping to leave behind the tribal disputes and civil wars of the Arabian peninsula. By the middle of the tenth century A.D., the once-unified Islamic empire had collapsed irretrievably into independent fragments, none of which provided the continuity over many generations that had made possible Alexandrian advances in geometrical astronomy from Apollonius to Hipparchus to Ptolemy.

The Arabs enjoyed both a relatively high level of science and a firsthand acquaintanceship with Ptolemy's *Almagest.* Yet many Islamic scientists were more interested in Aristotelian physical science. For them, Ptolemy's system was little more than a convenient computing device. In Spain in the twelfth century A.D., ibn-Rushd (Averröes, as he was known in the West) and al-Bitruji (Alpetragius) rejected Ptolemy's astronomy of epicycles, eccentrics, and equants for a system of concentric spheres more in accordance with Aristotelian physics. Averröes wrote that there was

> nothing in the mathematical sciences that would lead us to believe that eccentrics and epicycles exist. . . . The astronomer must, therefore, construct an astronomical system such that the celestial motions are yielded by it and that nothing that is from the standpoint of physics impossible is implied. . . . Ptolemy was unable to set astronomy on its true foundations. . . . The epicycle and the eccentric are impossible. We must, therefore apply ourselves to a new investigation concerning that genuine astronomy whose foundations are principles of physics. . . . Actually, in our time astronomy is nonexistent; what we have is something that fits calculation but does not agree with what is. (Duhem, *To Save the Phenomena*, 31).

And Alpetragius wrote:

> It is impossible to imagine numerous spheres with diverse motions for each planet, as Ptolemy assumed, or anything like it. . . . The assumptions and principles which he [Ptolemy] invented seemed to me a matter which I could not tolerate. I was not enthusiastic about his assumption—for example, that spheres eccentric to the center of the universe rotate about their eccenters, and that these centers rotate about other centers; that epicycles rotate about their centers . . . (Goldstein, *Al-Bitruji*, 59–60)

There would be little effort in Muslim Spain to develop further Ptolemy's geometrical astronomy. Nonetheless, the Islamic world did produce the most innovative addition to geometrical models of planetary motions achieved during the Middle Ages: the Tusi couple, a combination of uniform circular motions yielding net motion in a straight line.

In A.D. 1258 Mongol invaders under Hulagu Khan, a grandson of Genghis Khan, conquered Baghdad. Within a

Historical Perspectives on Contemporary Disputes

Details of al-Tusi's relationship with Hulagu Khan are not without interest even today, many hundreds of years after the fact. The allegation that al-Tusi, a Shiite Muslim, persuaded Hulagu to continue his attack on Baghdad and destroy the Sunni Caliphate contributes to current enmity between Shiite and Sunni Muslims. History has its uses and misuses. Sometimes it is exploited, and even distorted, for partisan purposes.

Other anecdotes raise the issue of government support for science and the compromises scientists sometimes must make. Al-Tusi had converted to the faith of his Sunni patrons. Then, following Hulagu's victory over the Sunni Caliphate, al-Tusi, in a self-serving recantation, described himself as having earlier fallen into the power of the heretics, and only now was he rescued from that place and ordered to observe the stars by Hulagu. Al-Tusi also found it expedient to rewrite introductions to a number of works in which he had lavishly praised his earlier benefactors.

Al-Tusi is not the only scientist ever pressured by his patron. With the recent demise of the Soviet Union and the opening of previously secret archives, an outpouring of scholarship is underway on the moral dilemmas and compromises of Soviet scientists. Similar studies have been undertaken of scientists working under Nazi domination. Even American scientists have complained of being pressured not to speak out against an administration's policies and not to release scientific studies whose conclusions are contrary to an administration's political inclinations.

year of his military triumph, Hulagu granted to Nasir al-din al-Tusi, an outstanding scholar formerly under the patronage of the dynasty conquered by Hulagu, his wish for an observatory.

Different sources attribute initiative for an observatory to al-Tusi, to Hulagu, and to Hulagu's brother, Mangu. Mangu Khan had a strong interest in mathematics and astronomy and may have asked Hulagu to send him al-Tusi to help build an observatory in his own capital city in China. Astrology apparently was behind Hulagu's interest in an observatory, though as expenses mounted, he is reported to have begun to doubt the utility of predicting immutable events if nothing could be done to circumvent them anyway.

Al-Tusi's observatory was constructed at Maragha, in northwest Persia (now Iran). Some of the most renowned scientists of the time, from as far as China to the east and Spain to the west, moved there, and the observatory's library was reported to have 400,000 volumes. Recent excavations, however, have found space for far fewer books.

Working with assistants and new instruments at Maragha, within a dozen years al-Tusi produced a new table of the planets' positions. He also wrote the *Al-Tadhkira*, a *Memoir on the Science of Astronomy*. In it he explained: "The scientific exposition that we wish to undertake will be a summary account of [astronomy] presented in narrative form. The details are expounded and proofs of the validity of most of them are furnished in the *Almagest*. Indeed, ours would not be a complete science if taken in isolation from the *Almagest* for it is a report of what is established therein" (Ragep, *Nasir al-Din al-Tusi's Memoir on Astronomy*, 19).

Al-Tusi choose not to go into the geometrical proofs available in the *Almagest*, but instead to concentrate on the physical situation: "These then are models and rules that should be known. We have only stated them here; their geometric proofs are given in the *Almagest*. Restricting oneself to cir-

Science and Islamic Culture

Funds from religious endowments helped finance the Maragha Observatory, which survived until at least A.D.1304 and possibly until A.D.1316, not only surviving Hulagu but spanning the reign of five or six more rulers as well. The administration of religious endowments often passed from father to son, and two of al-Tusi's sons succeeded him as director of the observatory.

Islamic traditionalists farther west, and thus free of the Mongol hegemony, criticized al-Tusi for transferring "the endowments of religious schools, mosques, and the hospices attached to them, making them his personal property . . . He also established schools for the heretics. . . . Finally he taught magic, for he was a sorcerer who worshipped idols" (Ragep, *Nasir al-Din al-Tusi's Memoir on Astronomy*, 19).

One historian has interpreted al-Tusi's diversion of religious endowments, normally devoted to institutions of charity and public assistance such as mosques, madrases (schools), and hospitals, to the operation of the observatory as an indication of the integration and harmonization of the observatory with Muslim culture and civilization. There were observatories at Maragha, Samarquand, Baghdad, and other places.

Not all was harmonious, however. The Istanbul observatory, built in A.D. 1577, would be torn down shortly after its completion, the attempt to pry into the secrets of nature suspected of having brought on misfortunes. In the wake of the famous comet of A.D. 1577, there had followed in quick order plague, defeats of Turkish armies, and the deaths of several important persons.

cles is sufficient in the entirety of this science for whoever studies the proofs. However, one who attempts to understand the principles of the motions must know the configuration of the bodies" (Ragep, *Nasir al-Din al-Tusi's Memoir on Astronomy*, 19).

The principle behind al-Tusi's innovative combination of uniform circular motions to produce motion in a straight line, the Tusi couple, is simply described. A straight-line motion can be produced by rolling (at any speed) a small circle within a large circle, with the planet fixed to the circumference of the rolling, smaller circle, provided that the diameter of the large circle is precisely twice the diameter of the small circle.

Al-Tusi's small circle, however, did not—indeed, could not—roll around in the large circle. Medieval Islamic astronomy aimed to produce a realistic physical model, not merely a mathematical formulation or fiction. Following Aristotle's physics, there could be no void in the heavens. Thus the large circle must be filled with some sort of celestial material, which would be torn apart by a small circle moving through it. Furthermore, after passage of the small circle, the celestial material would have had to be mended back together.

Instead, al-Tusi had both circles rotate, the small one twice as fast as the large one and in the opposite direction. This is mathematically equivalent to the small circle rolling at any speed in the large circle and avoids any possible collision of the planet and the small circle with celestial material filling the large circle.

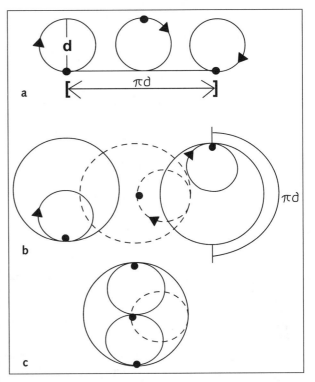

Figure 12.1: Straight Line Motion from Circular Motion. A combination of uniform circular motions can be devised to produce motion in a straight line.

a: First, imagine (above) a circle of diameter d rolling on a flat surface. A planet fixed to the circle moves halfway around the circle as the circle rolls a distance of half its circumference ($\pi d/2$), and completes a circuit as the circle rolls a distance πd.

b: Next (above), curve the flat surface into a circle of diameter D, with $D = 2d$. (The new circle is twice the size of the original circle.) Roll the small circle inside the large circle. A point on the small circle will come back into contact with the circumference of the large circle after one complete turn of the small circle. This occurs after the small circle moves a distance πd (its circumference) around the large circle, or half way around the large circle ($\pi d = \pi D/2$).

c: Finally, as the small circle rolls around the inside of the large circle, a point (planet) on the small circle constantly falls along a straight line, which is also a diameter of the large circle (in this case, the vertical diameter). Three positions of the small circle are shown above. Intermediate positions also place the planet on the same straight line. The net result is to convert uniform circular motion into seemingly straight-line motion.

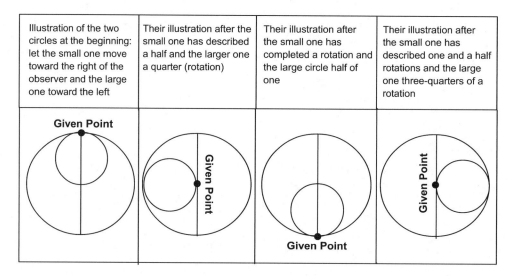

| Illustration of the two circles at the beginning: let the small one move toward the right of the observer and the large one toward the left | Their illustration after the small one has described a half and the larger one a quarter (rotation) | Their illustration after the small one has completed a rotation and the large circle half of one | Their illustration after the small one has described one and a half rotations and the large one three-quarters of a rotation |

Figure 12.2: A page from al-Tusi's *Al-Tadhkira*. Translation of figure captions by F. Jamil Ragep.

After formulating the version of his geometrical construction with circles, al-Tusi later created a version composed of spheres. Clearly, he was aiming at a physically plausible, Aristotelian cosmology with uniformly rotating spheres.

Al-Tusi's geometrical innovation commands attention in its own right. Furthermore, it is at the center of a historical mystery. Did Copernicus borrow this device for use in his own revolutionary astronomy?

It is not unknown in the history of science for identical problems and identical considerations to impel scientists in identical directions. General similarity is not conclusive proof that Copernicus borrowed a particular geometrical construction from an earlier source rather than that he independently discovered it. Geometrical drawings by al-Tusi and Copernicus, however, are strikingly similar, even sharing the same alphabetical letters (Arabic and Roman equivalents)!

Few Europeans could read Arabic, but Arabic and Persian astronomical writings were translated into Byzantine Greek and carried to Italy. The Vatican library also contains a Greek text of around A.D. 1300 on theoretical astronomy inspired by Islamic astronomy and incorporating the Tusi couple. Copernicus, in Rome in A.D.1500, might have seen it. Furthermore, other Greek and Latin materials using the Tusi couple were circulating in Italy when Copernicus studied there.

The likely connection between al-Tusi and Copernicus came to light at the end of the nineteenth century with the translation of sections of an A.D. 1389 manuscript of al-Tusi's book in the Bibliothèque Nationale in Paris. The astronomer-historian J.L.E. Dreyer, in his 1906 *History of the Planetary Systems from Thales to Kepler,* discussed al-Tusi's geometrical device and noted, somewhat cryptically and in a footnote: "Compare Copernicus, *De revolutionibus,* III, 4"

Mental Exercise: Rotating Circles to Rectilinear Oscillations

In the diagram below, determine the position of the planet at each eighth of a rotation if the equally sized circles rotate with:

1. equal speeds in opposite directions;
2. equal speeds in the same direction.

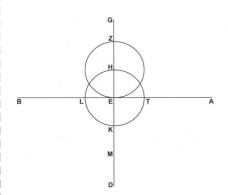

In addition to his famous Tusi couple, al-Tusi also devised a combination of equally sized rotating circles that produced a rectilinear oscillation. One circle's center rides on the circumference of the other circle, that circle rotating in the opposite direction and at half the speed of the first circle. In the above diagram, begin with the planet at Z carried around by circle ZE, whose center is H on circle HK. Rotate circle HK a quarter of a circle toward A, moving the center of the circle ZE from H to T (and the planet down onto line BA, beyond T toward A). Now rotate the circle (originally ZE) carrying the planet in the opposite direction and at twice the speed, completing half a circle of rotation. The planet, already rotated onto line BA, will move upward and then downward and in the direction of B, returning to the line BA. The center (originally H) of the circle (originally ZE) carrying the planet is now at T, placing the planet at E. Now repeat the process, rotating the center (originally H) of the circle (originally ZE) carrying the planet another quarter of a circle, from T to K. this moves the planet from E upward and to the right. Then the planet's circle rotates half a circle, carrying the planet to M. Repeating the process again, the center moves to L and the planet to E. When the center completes its rotation, returning to H, the planet moving at twice the speed completes two rotations, returning to Z. The net motion of the planet is up and down, and back and forth along the line GD between Z and M, with no motion right or left, toward A or B.

(p. 269). This suggestive note was generally ignored. Perhaps ethnocentric Western scholars were more comfortable with a generalization that, although Arabs had preserved Greek astronomy and transmitted it back to the West, they had contributed nothing to its advance.

Half a century after Dreyer's prescient but ignored footnote, a handful of surviving manuscripts of the work of the fourteenth-century Damascus astronomer Ibn al-Shatir were studied, and his lunar theory was recognized as being nearly identical with Copernicus's. Then a historical connection was forged from al-Shatir back to al-Tusi, and hence also a possible link between al-Tusi and Copernicus. Furthermore, both al-Shatir and Copernicus employed the Tusi couple to reproduce the motion of Mercury.

Traditionally, the origins of modern science have been traced to the Greeks. With the new realization of Arab contributions in astronomy, a few scholars are beginning to argue that dynamic revolutionary ideas in astronomy were developed in the Islamic world to refute the Greek astronomical tradition, and that the resulting Arabic mathematical techniques then made possible the astronomical revolution in the West associated with Copernicus and, more generally, the European Renaissance itself.

Claims of a dynamic revolution leading to the European Renaissance may go too far. A reasonable argument can also be advanced that the Tusi couple was yet another geometrical device to reproduce observed motions with combinations of uniform circular motions, as were deferents, epicycles, and equants. The Tusi couple, as clever as anything formulated by Apollonius, Hipparchus, or Ptolemy, resides comfortably within the ruling paradigm of Greek geometrical astronomy. We will need to look beyond Islamic planetary astronomy for additional causes of the scientific revolution of the sixteenth and seventeenth centuries.

Why did the scientific initiative shift from the Islamic world back to the West, as it had earlier shifted from the Greeks and Romans to the Arabs? Are there fundamental differences in attitudes toward science inherent in the natures of Islamic and Christian societies? Answers to these questions could encourage and direct current efforts to obtain the benefits of modern science for all the peoples inhabiting our planet.

13

REVIVAL IN THE WEST

Greek geometrical planetary astronomy was taken over, preserved, and improved in the Islamic world, even though the preponderant preference among Muslim philosophers was for a realistic model of the physical world composed of spheres rather than circles. In the West, where Ptolemy's *Almagest* was long unknown, interest focused even more strongly on Aristotelian physical science than on the astronomy of epicycles, deferents, and equants.

Physical and astronomical studies were customarily separate. Nonetheless, potential connections between geometrical planetary astronomy and Aristotle's physical cosmology warrant a brief survey of some particular developments in physics during the Middle Ages, that would be essential for the subsequent Copernican and Newtonian revolutions.

In Aristotle's physical world, the place of the Earth in the middle of the heavens was deducible from physical principles, and rotation of the Earth would lead to ridiculous consequences. Were a heliocentric hypothesis ever to receive serious attention, it would not only have to save the phenomena using uniform circular motions, but it would also have to be physically plausible. That could not occur absent modifications to Aristotle's physics. Ancient postulations of heliocentric hypotheses by both Hericlides and Aristarchus had been premature and easily ignored.

Because a Sun-centered system was eventually accepted, it is tempting to search for related changes in Aristotelian physics and then celebrate them as inevitable steps along the road to modern science. We might select evidence that supports a straightforward and logical development from *impressed force* to *momentum* to *inertia* and ignore all else. Retrospectively singling out part of an ancient scientist's work solely on the basis of its resemblance to current ideas is not a good way to practice history. But with limited space available and a definite issue at hand—the impossibility of a heliocentric system under Aristotelian physics, and yet its eventual acceptance—practical considerations suggest a focus on particular intellectual developments. This we do here, but not without first giving fair warning to readers.

Unlike Ptolemy's *Almagest,* produced after the handbook transmission period and consequently unavailable in the Roman world, Aristotle's writings were well known and the subject of extensive commentaries and criticisms. At least in that part of the Roman Empire in which learning and scholarship flourished, for as long as they did.

The Roman Empire split into eastern and western halves in A.D. 284, and briefly reunited under the emperor Constantine. He transferred his capital in A.D. 330 to Byzantium, which was renamed Constantinople (now Istanbul). By A.D. 476 the Western Empire had fallen to a series of invading barbarians. The Eastern Empire survived until A.D. 1453, when Constantinople fell to the Turks. The Eastern Empire enjoyed, during an otherwise slow decline, a brief renaissance in the sixth century A.D. Among the important scholarly works produced was a commentary on Aristotle by John Philoponus in A.D. 517. He is thought to have worked in Alexandria.

Aristotle had insisted on continuous contact between the mover and the body forced to move in a direction other than that of its natural motion. The need for continuous contact made movement in a vacuum supposedly impossible. It also made it difficult to explain how an arrow shot from a bow continued in motion with continuing contact between the mover and the moved. Somehow, air pushed forward by the arrow had to get around to the rear of the arrow and then push the arrow from behind. (*Antiperistasis* was Aristotle's term for this phenomenon.) Philoponus found this convoluted explanation bordering on the fantastic.

An alternative Aristotelian explanation also maintaining contact between the mover and the moved was that the air on being pushed received an impetus to motion and kept moving along with the arrow, remaining in contact with the arrow and pushing it, until the impetus impressed on the air dissipated. But this explanation was easily refuted. If the air were moving the arrow, then there would be no need for any contact between the arrow and the bow string, or between a hand and a thrown stone.

In place of the obviously inadequate Aristotelian explanations, Philoponus proposed instead a hypothesis of impressed force. He found it necessary to assume that some incorporeal motive force was imparted by the projector to the projectile. Furthermore, Philoponus's thoughts ranged upward from arrows and stones to stellar movements, which he also attributed to an impressed impetus. God who had created the Moon, the Sun, and the other heavenly bodies could also have implanted a motive power into them.

In A.D. 529 the emperor Justinian closed Plato's Academy in Athens, citing it as a center of paganism. Some of the persecuted scholars fled eastward, to Persia, taking with them the writings of Philoponus and others. There Aristotelian studies would flourish, especially under Islamic civilization beginning a century later. By the tenth century A.D., there was some Western contact with Muslim learning, and the eleventh and twelfth centuries saw an increase in translations of Arab and Greek works, mainly in Spain and in Sicily, on the borders between Islamic and Christian lands. Almost all of Aristotle's work, along with commentaries and criticisms, was available in Latin early in the thirteenth century.

Jean Buridan (ca. 1295–1358), who taught natural philosophy at the University of Paris, wrote several commentaries on books by Aristotle. The university curriculum was then based largely on the study of Aristotle's works on logic and natural philosophy, and a tightly integrated Aristotelian vision of the world dominated thinking. When Buridan raised questions about particular aspects of Aristotle's physics of motion, he undermined the entire foundation of Aristotelian cosmology.

Presentations were neither modern nor scientific in spirit. There was no genuine inquiry aimed at developing a new understanding of nature. Everyone knew what the questions were and what the answers would be. The aim of an exercise was skillful presentation of known information, not discovery of new information. In this, Buridan was a noteworthy exception.

He sought to answer whether a projectile after leaving the hand of the projector is moved by the air, or by what. This question he judged to be very difficult because Aristotle had not solved it well. Buridan, as Philopanus had earlier, rejected both Aristotelian explanations. Buridan argued that in the case of a toy top or a blacksmith's grinding wheel, which move for long times yet do not leave their places, air did not follow along and fill up the place of departure. Nor could air moving swiftly alongside the smith's wheel be the force moving it, because, if air on all sides near the wheel was cut off by a cloth, the wheel continued to move. Therefore it is not moved by the air.

Buridan proposed an alternative explanation that a motive force was impressed in the stone or other projectile. The impetus was continually decreased by air resistance and by the inclination of the projectile to move toward its natural place. Extending the concept of impetus to the Heavens, Buridan wrote:

> God, when He created the world, moved each of the celestial orbs as He pleased, and in moving them He impressed in them impetuses which moved them without his having to move them any more . . . And these impetuses which He impressed in the celestial bodies were not decreased nor corrupted afterwards, because there was no inclination of the celestial bodies for other movements. Nor was there resistance which would be corruptive or repressive of that impetus. (Hetherington, *Encyclopedia of Cosmology*, 50)

Although Buridan was passionate in his reasoning, he nonetheless rejected the Aristotelian position that there can be established necessary principles of physics. The problem was that truths necessary to philosophy could be contradictory to dogmas of the Christian faith. In A.D. 1270 the bishop of Paris condemned several propositions derived from the teachings of Aristotle, including the eternity of the world and the necessary control of terrestrial events by celestial bodies. In A.D. 1277 the Pope directed the bishop to investigate intellectual controversies at the university, and within three weeks the bishop condemned over two hundred more propositions. Excommunication was the penalty for holding even one of the damned errors.

Buridan helped develop a *nominalist* thesis. It conceded the divine omnipotence of Christian doctrine but at the same time established natural philosophy as a respected subject and defined it in a way that, while acceding to religious authority, also removed natural philosophy from under religious authority. Science was understood as a working hypothesis in agreement with observed phenomena. But we cannot, Buridan argued, insist on the truth of any particular working hypothesis. God could have made the world in some different manner but with the same set of observational consequences. Therefore, scientific theories are tentative, not necessary, and can pose no challenge to religious authority.

Nominalism and Instrumentalism

Nominalism has much in common with *instrumentalism*. Both philosophical concepts posit scientific theories as working hypotheses with no necessary links to reality.

One might speculate that Catholic historians and philosophers of science in the twentieth century, justifiably impressed with fourteenth-century nominalism and also eager to praise the admirable achievements of that era by scholars working within and supported by the Catholic church, consequently were predisposed to formulate the concept of instrumentalism. Nominalism stripped of its religious context became instrumentalism, and nominalists were favorably pictured as forerunners of modern, philosophically sophisticated instrumentalists.

The nominalist thesis was a convenient stance to take in a time when religious matters were taken seriously and heretical opinions could place their adherents in serious trouble with powerful ecclesiastical authorities. But we should not jump to the conclusion that Buridan was playing a cynical game when he ended his presentation of impetus theory by deferring to religious authority. He wrote: "But this I do not say assertively, but so that I might seek from the theological masters what they might teach me in these matters as to how these things take place" (Hetherington, *Encyclopedia of Cosmology*, 50). Buridan almost certainly was sincere in this statement.

In the new intellectual climate of nominalism, imaginative and ingenious discussions, even concerning the possible rotation of the Earth, flourished. Some historians assert that the scientific revolution of the sixteenth and seventeenth centuries owes much to the condemnation of 1277. Though intended to contain and control scientific inquiry, the condemnation may have helped free cosmology from Aristotelian prejudices and modes of argument.

But if so, why did scholars wait hundreds of years before repudiating Aristotelian cosmology? Hypothetical cosmologies are not the stuff of revolution. Not until the goal of "saving the appearances," as the nominalist endeavor has been called, was replaced with a quest to discover physical reality was Aristotelian cosmology destroyed, to be replaced with a new worldview. Confidence that the essential structure and operation of the cosmos is knowable was a prerequisite to the work of Copernicus, Galileo, Kepler, and

Newton. Inevitably, some of their necessary cosmological principles would conflict with dogmas of the Christian faith.

Buridan's thoughts were further advanced by his student Nicole Oresme. Regarding the possible rotation of the Earth, Oresme wrote: "But it seems to me, subject to correction, that one could support well and give luster to the opinion that the Earth, and not the Heavens, is moved with a daily movement" (Hetherington, *Encyclopedia of Cosmology*, 451).

Oresme began by recounting observations and arguments raised against the possible rotation of the Earth. One, we see the Sun, the Moon, and the stars rise and set from day to day. Two, there is no wind blowing continuously from the east, as there would be were the Earth rotating swiftly from west to east. And three, an object thrown upward falls straight down. Oresme elaborated:

> The third experience is that which Ptolemy advances: If a person were on a ship moved rapidly eastward and an arrow were shot directly upward, it ought not to fall on the ship but a good distance westward away from the ship. Similarly, if the Earth is moved so very swiftly in turning from west to east, and it has been posited that one throws a stone directly above, then it ought to fall, not on the place from which it left, but rather a good distance to the west. But in fact the contrary is apparent. (Hetherington, *Encyclopedia of Cosmology*, 451)

Oresme proceeded to demolish the arguments against the possible rotation of the Earth. Yes, the Sun, the Moon, and the stars are perceived to move with respect to the Earth, but a person on a moving ship would similarly perceive a ship at rest as moving. There was no strong wind from the east because the lower air and the water are moved with the Earth, as is the air enclosed in the cabin of a moving ship. And the arrow or stone projected upward was also moved eastward with the diurnal movement and thus returned to the same place on the Earth from which it left. Oresme summed up his arguments, emphasizing the concept of relative motion: "It is apparent then how one cannot demonstrate by any experience whatever that the Heavens are moved with daily movement, because . . . if an observer is in the Heavens and he sees the Earth clearly, the Earth would seem to be moved; and if the observer were on the Earth, the Heavens would seem to be moved" (Hetherington, *Encyclopedia of Cosmology*, 452).

Oresme acknowledged that the rotation of the Earth seemed to be against holy scripture, which said that the Sun riseth, and goeth down, and returneth to his place; and God created the orb of the Earth, which will not be moved; and the Sun was halted in the time of Joshua and turned back in the time of King Ezechias. To these objections Oresme answered that holy scripture was not to be read literally but was to be understood, rather, as conforming to the manner of common human speech.

Furthermore, there was an aesthetic reason to place the observed rotation in the Earth rather than in the heavens. Oresme noted that: "Since all the effects which we see can be accomplished and all the appearances saved by substituting for this [the rotation of the heavens] a small operation, the diur-

Figure 13.1: Ptolemy and Regiomontanus are seated beneath an armillary sphere. It was used to measure coordinates (longitude and latitude) of celestial bodies. Frontispiece, Regiomontanus, *Epitome of the Almagest*, 1496. Image copyright History of Science Collections, University of Oklahoma Libraries.

nal movement of the Earth, which is very small in comparison with the Heavens, and without making the operations so diverse and outrageously great, it follows that if the Heaven rather than the Earth is moved, God and Nature would have made and ordained things for nought, but this is not fitting" (Hetherington, *Encyclopedia of Cosmology*, 452).

Oresme, however, could not leave things here. He was too close to asserting truths necessary to philosophy that could be contradictory to dogmas of the Christian faith. He concluded: "Yet nevertheless everyone holds and I believe that the Heavens and not the Earth are so moved . . . not withstanding the arguments to the contrary . . . because they are 'persuasions' which do not make the conclusions evident. This, that which I have said by way of diversion in this manner can be valuable to refute and check those who would impugn our faith by argument" (Hetherington, *Encyclopedia of Cosmology*, 453). If Oresme the nominalist was not ready to insist on the truth of the heliocentric hypothesis, he nonetheless had worked out much of the reasoning that more intrepid philosophers would employ centuries later.

Aristotelianism and Christian theology fused into *Scholasticism*, a way of thought which permeated Western Europe between roughly A.D. 1200 and 1500, especially in universities. Unsympathetic critics might consider much of Scholastic philosophy merely semantic squabbling. Indeed, the adjective *scholastic* has become pejorative, defining obscure and irrelevant discussion of no practical value.

Scholasticism retained a hold on teaching in Catholic seminaries but lost ground in the larger world to humanism. From roughly A.D. 1300 to 1600, humanism overlapped Scholasticism. With humanism came a renewed interest in Plato. *Neoplatonism*, also called *Neopythagoreanism*, included a new belief in the possibility and importance of discovering simple arithmetical and

Mirroring and Echoing the Divine Universe

The theologian Augustine of Hippo (A.D. 354–430) believed that the essence of beauty lies in resemblance to the divine universe. Seemingly discordant elements are brought into ultimate harmony under a single, overarching geometrical order.

Beautiful music is composed of simple arithmetic ratios between musical notes. The result echoes the divine universe, which consists of similar ratios. The medieval motet, an unaccompanied choral composition with different texts (sometimes in different languages!) sung simultaneously over a Gregorian chant fragment, originated in the thirteenth century. It was sung as part of church services, for the greater glory of God and for the enlightenment, instruction, and diversion of man. The music's intricate polyphonic form, in which every little detail has its place and meaning, supposedly echoed the superbly organized divine universe. Perhaps the greatest expression of this musical genre is found in the eighteenth-century fugues of Johann Sebastian Bach, Kapellmeister and Director Chori Musici in Leipzig.

Resemblance to the divine universe as the essence of beauty also applies to architecture, especially to Gothic cathedrals, constructed between the twelfth and sixteenth centuries. Dedicating the new choir of the abbey church of Saint-Denis in 1144, Abbot Suger called it the embodiment of the mystical vision of harmony that divine reason had established throughout the cosmos. The geometrical regularity and harmony of Gothic cathedrals was, he believed, as literal a depiction of a spiritual ideal as could be built in stone. Architecture mirrored the divine universe. Some Western medieval architects were so convinced of the power of geometrical regularity to stabilize structures that when the Milan cathedral showed signs in the 1390s of collapsing, one solution proposed was to *increase* its height, in order to form one section into a perfect square. Fortunately, instead of trying to bring the cathedral into closer harmony with the divine universe, more buttresses were added. Other cathedrals, however, did collapse, including Worcester in 1175, Lincoln in 1240, Beauvais—the tallest Gothic cathedral ever built—in 1282, Ely in 1321, and Norwich in 1361.

Painters, too, have consciously attempted to reflect the harmony of the cosmos. The squares and primary colors favored by the twentieth-century Dutch painter Piet Mondrian are one example.

In the Islamic world, in contrast, the Koran, the sacred text of Islam, is the unadulterated word of God (Allah), and Arabic script is the means for transmission and visualization of the divine message. Idolatry (in any form other than calligraphy) is discouraged, and figural images are not found in Islamic religious architecture, such as mosques and shrines.

Glenn Seaborg, the Nobel Prize–winning scientist who discovered 10 atomic elements, including plutonium and seaborgium, noted that the educated person of today and tomorrow can no more ignore science than his predecessors of the Middle Ages could ignore the Christian church or the feudal system. It now appears that neither they nor we can disentangle cathedrals and their music and architecture from science.

geometrical regularities in nature, and a new view of the Sun as the source of all vital principles and forces in the universe.

A re-naissance is a re-birth, and the initial thought of Renaissance humanists was that they were facilitating a rebirth of Greek philosophy and values. Today, few persons would look to the past for knowledge from a higher civilization, but people growing up in the ruins of the Greek Parthenon or the Roman Colosseum understandably might have looked longingly back to Greek and Roman civilizations. The recovery, translation, and diffu-

Figure 13.2: Spherical Scheme of the Universe from Petrus Apianus, *Cosmographicus Liber*, 1540. First published in 1524, this popular introduction to astronomy, geography, cartography, surveying, navigation, weather and climate, the shape of the Earth, map projections, and mathematical instruments was reprinted throughout the sixteenth century. The woodcut shows a central spherical Earth surrounded by solid spheres carrying the Moon, Mercury, Venus, Sun, Mars, Jupiter, Saturn, and the fixed stars. Born Peter Bienewitz, he Latinized his name to Petrus Apianus (*biene* is German for "bee," which is *apis* in Latin). Apianus, who lived from 1495 to 1552, was professor of mathematics in Ingolstadt (north of Munich) and a favorite of Charles V, the Holy Roman Emperor and King of Aragon and of Castile (now Spain). Image copyright History of Science Collections, University of Oklahoma Libraries.

sion of lost classical works marked the first stage of the Renaissance and humanism. The Renaissance began in Italy in the fourteenth century A.D. and spread to universities north of the Alps in the fifteenth and sixteenth centuries A.D.

Inconsistencies within individual ancient works and between different authors, and discrepancies in the sciences between classical theory and contemporary observation, initially could be attributed to defects in transmission and translation. Eventually, however, critical thought was stimulated, and what had begun as a rebirth or recovery of old knowledge mutated into the creation of new knowledge.

Ptolemy's *Almagest* became available to scholars in the Latin-reading world late in the fifteenth century. In 1460 a representative of the Pope arrived in Vienna seeking aid for a crusade against the Turks to recapture Constantinople. He also enlisted Georg Peurbach (1423–1461) and his pupil Johannes Müller (1436–1476), known as Regiomontanus (after his home city of Königsberg: *König*, "king or regent" [i.e., *regio*]; and *berg*, "mountain" [i.e., *montanus*]), to prepare a new translation of the *Almagest*. Peurbach died the next year, but not before extracting a promise from Regiomontanus to see the work through to completion. Regiomontanus would die in 1476, perhaps of the plague, or perhaps murdered by sons of a scholar he had criticized. Regiomontanus's *Epitome of the Almagest*, completed in 1463, was first printed in 1496 in Venice, and reprinted in Basel in 1543 and in Nuremberg in 1550. A translation of the *Almagest* from Arabic was printed in 1515, and a 1451 translation from Greek to Latin finally appeared in print in 1528.

Copernicus would rely heavily on the rebirth of Ptolemy's mathematical astronomy, as midwifed by Peurbach and Regiomontanus, for both its geometrical techniques and its philosophical human values. Scholastic discussions of the possibility of a rotating Earth and humanism's general emphasis on Platonic thought, including in particular the Neoplatonic view of the Sun as the source of all vital principles and forces in the universe, were also elements in Copernicus's intellectual world.

14

COPERNICUS AND PLANETARY MOTIONS

A great underlying theme dominating Western astronomy from Plato and the Greeks to the end of the Middle Ages and into the early Renaissance was a continuing and evolving struggle to represent observed planetary motions as combinations of uniform circular motions. Copernicus's *De revolutionibus orbium caelestium* (*On the Revolutions of the Celestial Spheres*) of 1543 is as full of epicycles and deferents as was Ptolemy's *Almagest* fourteen centuries earlier. Scholars have quibbled over who had more circles, Copernicus or Ptolemy, but no one denies a plethora of circles in each of their books. Copernican astronomy was the culmination of Greek geometrical astronomy and uniform circular motion.

In what way, then, was Copernican astronomy a revolutionary departure from the astronomy that preceded it, if indeed it was? Thomas Kuhn has argued that *De revolutionibus,* while a *revolution-making* book, was not a *revolutionary* book. It was almost entirely within an ancient astronomical tradition. Yet it contained a few novelties that would lead to a scientific revolution unforeseen by its author.

Another question, whether or not we label Copernicus a revolutionary, is what made him so dissatisfied with Ptolemy's planetary model that he was motivated to devise a new theory. In modern science, the usual suspect for the role of crucial anomaly resisting conformity with theory is a new observation. In Copernicus's time, however, no startling new astronomical observations refuted Ptolemaic astronomy. Nor did Copernicus's heliocentric system provide a better match between theory and observation than had Ptolemy's geocentric system. Copernicus, himself, acknowledged that Ptolemy's planetary theory was consistent with numerical data. What, then, demanded change? On what basis other than observation might one theory be judged better than its rival?

To answer these questions, we must look to the paradigm or themata informing Copernicus about nature and how it behaved. His historical, cultural, and sociological settings—not least the education he received and the philosophical and aesthetic values he absorbed—are important facets of a history of Copernican astronomy. What were his underlying beliefs and worldviews—scientific and otherwise? Quasi-aesthetic choices based on them, not on observation, would guide Copernicus at critical junctions along his path toward change, possibly revolutionary change and certainly revolution-making change.

Nicolaus Copernicus was born in 1473 in the city of Torun, where a major trade route crossed the Vistula River. The Teutonic Knights, a group of German military monks, had built a castle in Torun in the thirteenth century and ruled over the city. Seven centuries later, Nazi historians attempted to claim Copernicus for the glory of the Third Reich, even though the people of Torun had revolted against the Teutonic Knights in 1454 and Torun had become part of Poland, only to be annexed to Prussia in 1793 and reclaimed by Poland in 1918.

Copernicus's father was a merchant and a leading citizen of Torun. Copernicus's mother came from an even more eminent and wealthy family. Her brother, Lucas Watzenrode, became guardian of Copernicus and his siblings in 1483, when Copernicus's father died.

Watzenrode had attended Cracow University in Poland and the University of Bologna in Italy, where he studied church law. Probably in existence as early as the 1080s, and chartered by the Holy Roman Emperor Frederick Barbarossa in 1158, Bologna is the oldest continuing university in the Western world. Watzenrode obtained a position at Frombork (Frauenburg) Cathedral, built by the Teutonic Knights in the fourteenth century but under Polish influence since 1466. Watzenrode became both bishop and ruler of Warmia (Ermland), including the Cathedral church at Frombork. Warmia was a sort of vassal state of Poland, though the Teutonic Knights still owned much of its land. Squeezed uncomfortably between Poland and the Knights, the bishop's primary task was simply to survive.

In 1491 Copernicus was admitted to the University of Cracow. The city, astride a major trade route from Europe to the East, was famous for its wealth and culture. The university, first chartered in 1364, was one of the first northern European schools to teach Renaissance humanism, which had originated in Italy. Astronomy at Cracow, however, still was taught largely in terms of Aristotelian physics. Several mathematics and astronomy books collected and annotated by Copernicus while he was at Cracow furnish evidence of his early interest in astronomy.

Watzenrode intended to appoint Copernicus to a position at Frombork Cathedral after a few years at Cracow University. Waiting for the next open position, in 1496 Copernicus was sent to Bologna to study church law. The astronomy professor there, Domenico Novara, was one of the leaders in the revival of Platonic and Pythagorean thought and Greek geometrical astronomy, and we will find this intellectual heritage on prominent display half a century later in Copernicus's *De revolutionibus*. Teacher and student probably discussed Ptolemaic astronomy. Copernicus made observations, recording in his notebook the exact time that the Moon passed in front of a bright star.

In 1497 Watzenrode appointed Copernicus a canon at Frombork Cathedral. Copernicus, however, remained in Italy for the great Jubilee of 1500, when pilgrims traveled to Rome to seek pardon for their sins. He returned to Frombork in 1501, only long enough to receive permission to complete his studies in Italy. It was not uncommon then to begin at one university and finish at another. This Copernicus did, completing his legal studies at the University of Padua in 1503. He chose to take the final examination and graduate at the University of Ferrara, perhaps to avoid the cost of a celebration for his many friends in Padua. Copernicus returned to Padua and studied medicine there before returning to Poland for good, probably early in 1506.

During his first six years back in Poland, Copernicus, rather than taking up his duties at Frombork, waited on his uncle as medical attendant and also helped him govern Warmia. Copernicus administered taxes, dispensed justice, and traveled often on government business and as a diplomat. When Watzenrode died in 1512, Copernicus finally took up his long-delayed duties at Frombork Cathedral. The Teutonic Knights had built a fortified wall around the cathedral, and Copernicus lived in one of its towers. The tower opened onto the wall, from which Copernicus could make observations, and one of the first records of his stay at Frombork is an observation of the planet Mars. Observations, however, did not turn Copernicus into a revolutionary.

In 1514 the pope convened a general council in Rome to study problems involving calendars, especially the date of Easter. Copernicus was invited but declined to attend, replying that before the calendar problem could be resolved the theory of the motions of the Sun and the Moon needed to be better known, and that he was working on this matter. If we insist on an explanation involving observation for the Copernican revolution, the calendar problem is the best bet. Anomaly led to a sense of crisis, which encouraged Copernicus to work on the problem, and his employer or patron (the Church) to allow him time for this important work.

As much, however, as it may have motivated Copernicus to study astronomy and planetary motions, the calendar problem did not suggest to him any particular scientific solution. He might as well have attempted to improve Ptolemy's model as to replace it with a new theory. Instead, as we will find in Copernicus's own words, philosophical opinions and aesthetic values underlay his dissatisfaction with Ptolemy's planetary model and helped guide him toward a new theory.

Around this same time, Copernicus was distributing handwritten copies of a sketch of his hypotheses of the heavenly motions. It is probably the manuscript of six sheets of paper expounding the theory that the Earth moves while the Sun stands still that was listed in 1514 by a Cracow professor cataloging his books. The *Commentariolus* (*Little Commentary*) soon disappeared from circulation, and wasn't published until 1878 after a handwritten copy was found in Vienna.

The Calendar Problem

The calendar used in Copernicus's time was that dictated by Julius Caesar in the first century B.C. It was linked to the Sun rather than to the Moon. Its year was, and still is, 365 days long, with an extra day added to the month of February every fourth year. The actual length of the year, however, is about 11 minutes less than Caesar decreed. This works out to an error of one whole day every 128 years.

The Church linked Easter not to the Sun but to the Moon. Easter occurs on the first Sunday after the first full moon on or after the vernal equinox.

Translating a fixed date from a lunar calendar to a variable date in a solar calendar containing an error of several days over several centuries is not easy. Embarrassingly, the Church was unable to predetermine very far in advance the Julian date of Easter.

The Gregorian calendar, based on computations using the Copernican system, was introduced in 1582 by Pope Gregory VIII. No longer is a leap day added to years divisible by 100, unless they are also divisible by 400. Thus 1900 was not a leap year, but 2000 was.

Copernicus began by noting that ancient astronomers had assumed that heavenly bodies moved uniformly and had attempted to explain apparent motions by the principle of regularity. Callippus and Eudoxus had tried with concentric spheres but failed. Most scholars then accepted eccentrics and epicycles.

Ptolemy's planetary theory was consistent with numerical data but contained equants: planets moved neither on a deferent nor about the center of an epicycle. To Copernicus: "A system of this sort seemed neither sufficiently absolute nor sufficiently pleasing to the mind" (*Commentariolus*, 57). What Copernicus meant by this will become more apparent in *De revolutionibus*. The point to be emphasized here is that Copernicus acknowledged agreement of Ptolemy's theory with observation but objected to Ptolemy's employment of the equant point on aesthetic or philosophical grounds—it was not pleasing to Copernicus's mind.

He sought "a more reasonable arrangement of circles" in agreement with all observed apparent motions "and in which everything would move uniformly about its proper center" (*Commentariolus*, 47–48). In such an arrangement, the center of the Earth was not the center of the universe. Rather, all the spheres (except the Moon's) revolved about the Sun as their midpoint, and therefore the Sun was the center of the universe. Rotation of the Earth accounted for the apparent daily rotation of the stars; the apparent movements of the Sun were caused by the Earth revolving around the Sun; and the apparent retrograde motions of the planets were caused by the motion of the Earth. Copernicus concluded: "The motion of the Earth, alone, therefore, suffices to explain so many apparent inequalities in the heavens" (*Commentariolus*, 59). He provided sizes for orbits, their angles of inclinations, periods of revolution, and latitudes and longitudes for a few planetary alignments; but detailed mathematical demonstrations were reserved for his larger work.

Copernicus had shown that a heliocentric system might potentially be constructed on his principles, but much more needed to be done merely to match, let alone improve on, Ptolemy's already fully developed geocentric system. From his *Commentariolus* and its simple Sun-centered system of planets moving in circles around the central body, Copernicus's planetary model would evolve over the years between 1514 and 1543 into the complicated mathematical machinery of epicycles and deferents found in his *De revolutionibus*.

One more assumption appeared in the *Commentariolus*. Copernicus stated that "the distance from the Earth to the Sun is imperceptible in comparison with the height of the firmament" (*Commentariolus*, 58). A universe vastly larger than the region of the Sun and Earth would be a major part of the Copernican revolution away from the closed medieval universe and the central position of human beings, but there is no hint of these matters here. Nor is there any hint of the future controversy over the absence of an observed stellar parallax, predicted from Copernicus's heliocentric theory but too small to detect were the stars at great distances from the Earth. Copernicus explored neither of these issues implicit in his new planetary model.

In 1516 Copernicus was given the task of managing two outlying estates, and he moved to Allenstein Castle, about 50 miles from Frombork. War broke out between the Teutonic Knights and the Poles in 1519. During 1520, while the town of Frombork burned, Copernicus was safely behind the cathedral wall, with time to work on his book on a new system of the world. During 1521 he was besieged in Allenstein and again had time for astronomy. In the peace that followed, Copernicus was busily employed in repairing the damage from years of war. Many church officials had fled Warmia, and during 1523 Copernicus acted as administrator-general of the tiny country.

Copernicus's hypothesis concerning the heavenly motions was receiving attention. It was the subject of a lecture given before the pope in 1533, and in 1536 a cardinal wrote from Rome to ask Copernicus for details of his system. Attention was not all positive, however; a satirical play in 1531 made fun of Copernicus. More seriously, Martin Luther criticized

> the new astronomer, who wants to prove that the Earth goes round, and not the heavens, the Sun, and the Moon; just as if someone sitting in a moving wagon or ship were to suppose that he were at rest, and that the Earth and the trees were going past him. But that is the way nowadays; whoever wants to be clever must needs produce something of his own, which is bound to be the best because he has produced it! The fool will turn the whole science of Astronomy upside down. But, as the Holy Writ declares, it was the Sun and not the Earth which Joshua commanded to stand still. (Kuhn, *Copernican Revolution*, 191)

Luther had initiated the Reformation early in the sixteenth century. He accused the papacy of departing from the teachings of the Bible and insisted that Church officials and their interpretations should not be allowed to come between believers and the Bible. Literal adherence to the Bible was thus the foundation of the Protestant revolt against Catholic religious hegemony, and Protestants consequently rejected metaphorical and allegorical interpretations. Therefore, it was Protestants, not Catholics, who initially found the heliocentric theory unacceptable, especially because of Joshua's command to the Earth to stand still (Joshua 10:13). Prior to the Counter Reformation, in whose tendrils Galileo would become entangled, the Catholic Church was more liberal in its interpretation of the Bible and thus more accepting of Copernican astronomy than were Protestants. Copernican astronomy would be taught in some Catholic universities and would be used for the new calendar promulgated by the pope in 1582.

In 1539 a young German professor of mathematics, known as Rheticus, after his birthplace, Rhaetia, showed up at Frombork to learn more about Copernicus's system. It was a brave action. Rheticus could have found himself in disfavor with authorities back home at his Protestant university, or he might have been persecuted as a Protestant in Catholic Warmia. Copernicus, too, displayed courage, daring to host a heretic. Rheticus wrote a biography of Copernicus, unfortunately lost.

Copernicus was almost finished with his *De revolutionibus,* and he generously shared it with Rheticus. In 1540 Rheticus sent to a printer the *Narratio prima,* or *First Account,* of part of Copernicus's theory. Rheticus may have planned a second account, and perhaps more, but he was forestalled by the publication of Copernicus's own book in 1543. Perhaps Copernicus was encouraged to publish by the favorable reception given Rheticus's *Narratio prima.*

It was necessary to abandon the hypotheses of the ancient astronomers, Rheticus wrote. The motion of the Earth could produce most of the apparent motions of the planets. Furthermore, only in the heliocentric theory "could all the circles in the universe be satisfactorily made to revolve uniformly and regularly about their own centers, and not about other centers—an essential property of circular motion" (*Narratio prima,* 137). Here he was criticizing equant points as a violation of uniform circular motion.

Another point of criticism was that early planetary models were deficient in harmony (or unity or symmetry or interconnection). Ancient astronomers had

fashioned their theories and devices for correcting the motion of the heavenly bodies with too little regard for the rule which reminds us that the order and motions of the heavenly spheres agree in an absolute system. . . . We should have wished them, in establishing the harmony of the motions, to imitate the musicians who, when one string has either tightened or loosened, with great care and skill regulate and adjust the tones of all the other strings, until all together produce the desired harmony, and no dissonance is heard in any. . . .

Under the commonly accepted principles of astronomy, it could be seen that all the celestial phenomena conform to the mean motion of the Sun and that the entire harmony of the celestial motions is established and preserved under its control. . . .

. . . Hence the mean motion of the Sun would necessarily be perceived in all the motions and appearances of the other planets in a definite manner, as appears in each of them. Thus the assumption of the motion of the Earth on an eccentric provides a sure theory of celestial phenomena, in which no change should be made without at the same time re-establishing the entire system, as would be fitting . . .

. . . the remarkable symmetry and interconnection of the motions and spheres, as maintained by the assumption of the foregoing hypotheses, are not unworthy of God's workmanship and not unsuited to these divine bodies. (*Narratio prima*, 138–40, 145)

In Ptolemy's system, the parts were arbitrarily arranged. Theory for one planet's motion was independent of theories for other planets. In Copernicus's system, the parts had a natural and necessary order. Were one circle changed, all the others would have to be adjusted, too.

In the intellectual climate of Renaissance humanism, the appeal to coherence and mathematical harmony must have resonated strongly among Neoplatonists and Neopythagoreans. Rheticus made the appeal explicit: "Following Plato and the Pythagoreans, the greatest mathematicians of that divine age, my teacher [Copernicus] thought that in order to determine the causes of the phenomena, circular motions must be ascribed to the spherical Earth" (*Narratio prima*, 147–48). Commenting on the fact that in the Copernican system there were six moving spheres revolving around the Sun, the center of the universe, Rheticus asked rhetorically:

Who could have chosen a more suitable and more appropriate number than six? By what number could anyone more easily have persuaded mankind that the whole universe was divided into spheres by God the Author and Creator of the world? For the number six is honored beyond all others in the sacred prophecies of God and by the Pythagoreans and the other philosophers. What is more agreeable to God's handiwork than that this first and most perfect work should be summed up in this first and most perfect number? Moreover, the celestial harmony is achieved by the six aforementioned movable spheres. For they are all so arranged that no immense interval is left between one and another; and each, geometrically defined, so maintains its position that if you should try to move any one at all from its place, you would thereby disrupt the entire system. (*Narratio prima*, 147)

God, Plato, Copernicus, and the study of astronomy were all linked together. Rheticus wrote:

Moreover, divine Plato, master of wisdom . . . affirms not indistinctly . . . that astronomy was discovered under the guidance of God. . . . [Copernicus was] seeking the mutual relationship which harmonizes them all [observations] . . . not indeed without divine inspiration and the favor of the gods. . . . (*Narratio prima*, 163)

Platonic and Pythagorean resonances are also sounded in Copernicus's *De revolutionibus* of 1543.

De revolutionibus began with a prefatory letter to the pope. Copernicus was well aware

that certain people, as soon as they hear that in this book about the Revolutions of the Spheres of the Universe I ascribe movement to the earthly globe, will cry out that, holding such views, I should at once be hissed off the stage. . . .

NICOLAVS COPERNICVS
Mathematicus.

Quid tum? si mihi terra mouetur, Solq, qui escit,
Ac cœlum: constat calculus inde meus.

M.   D.   XLI.

Figure 14.1: Portrait of Copernicus. Woodcut by Tobias Stimmer, 1451. Stimmer also painted this portrait of Copernicus in his decoration for the famous astronomical clock in the Strasbourg Cathedral, and wrote that it was a true likeness from Copernicus's own self-portrait. The lily of the valley in Copernicus's left hand is the standard icon for a medical doctor. Image copyright History of Science Collections, University of Oklahoma Libraries.

Thinking therefore within myself that to ascribe movement to the Earth must indeed seem an absurd performance . . . I hesitated long whether, on the one hand, I should give to the light these my Commentaries written to prove the Earth's motion, or whether, on the other hand, it were better to follow the example of the Pythagoreans and others who were wont to impart their philosophic mysteries only to intimates and friends, and then not in writing but by word of mouth. . . . (*De revolutionibus*, preface)

Copernicus anticipated that the pope would want to know how he "came to dare to conceive such motion of the Earth, contrary to the received opinion of the Mathematicians and indeed contrary to the impression of the senses" (*De revolutionibus*, preface). He had been induced to think of a new system because:

First, the mathematicians are so unsure of the movements of the Sun and Moon that they cannot even explain or observe the constant length of the seasonal year. Secondly, in determining the motions of these and of the other five planets, they use neither the same principles and hypotheses nor the same demonstrations of the apparent motions and revolutions. So some use only homocentric circles, while others [use] eccentrics and epicycles. Yet even by these means they do not completely attain their end. Those who have relied on homocentrics, though they have proven that some different motions can be compounded there from, have not thereby been able fully to establish a system which agrees with the phenomena. Those again who have devised eccentric systems, though they appear to have well-nigh established the seeming motions by calculations agreeable to their assumptions, have yet made many admissions [the equant] which seem to violate the first principle of uniformity in motion. Nor have they been able thereby to discern or deduce the principle thing—namely the shape of the Universe and the unchangeable symmetry of its parts. With them it is as though an artist were to gather the hands, feet, head, and other members for his images from diverse models, each part excellently drawn, but not related to a single body, and since they in no way match each other, the result would be monster rather than man. So in the course of their exposition, which the mathematicians call their system, . . . we find that they have either omitted some indispensable detail or introduced something foreign and wholly irrelevant. (*De revolutionibus*, preface)

The principal thing was the shape of the universe and the unchangeable symmetry of its parts. What Copernicus meant by this is not immediately evident. He promised "though my present assertions are obscure, they will be made clear in due course" (*De revolutionibus*, preface).

How had Copernicus, discontented with the old system, come upon a new system? He explained to the pope:

I pondered long upon this uncertainty of mathematical tradition in establishing the motions of the system of the spheres. At last I began to chafe that philosophers could by no means agree on any one certain theory of the mechanism of the Universe, wrought for us by a supremely good and orderly Creator, though in other respects they investigated with meticulous care the minutest points relating to its circles. I therefore took pains to read again the works of all the philosophers on whom I could lay hand to seek out whether any of them had ever supposed that the motions of the spheres were other than those demanded by the mathematical schools. (*De revolutionibus*, preface)

It was customary early in the Renaissance to look to the past for knowledge. It was also typical to link *orderly* and *good* and assume that the universe was so constructed.

Some ancient philosophers had supposed different motions of the spheres, according to the Roman writer Plutarch, whom Copernicus quoted: "Philolaus the Pythagorean says that she [the Earth] moves around the [central] fire on an oblique circle like the Sun and Moon. Heraclides of Pontus and Ecphantus the Pythagorean also make the Earth to move, not indeed through space but by rotating around her own center as a wheel on an axle from West to East" (*De revolutionibus*, preface).

Emboldened by ancient example and precedent, Copernicus began to consider the mobility of the Earth. He wrote:

I have at last discovered that if the motions of the rest of the planets be brought into relation with the circulation of the Earth and be reckoned in proportion to the circles [orbits] of each planet, not only do their phenomena presently ensue, but the orders and magnitudes of all stars and spheres, nay the heavens themselves, become so bound together that nothing in any part thereof could be moved from its place without producing confusion of all the other parts and of the Universe as a whole. . . . (*De revolutionibus*, preface)

Copernicus concluded his prefatory letter to the pope with a wish that his labors might contribute to the Church, especially to reform of the ecclesiastical calendar. Even in his remote corner of the world, Copernicus wrote flatteringly, the pope was regarded as the most eminent by virtue of the dignity of his office and of his love of letters and science. By his influence the pope (so Copernicus hoped) might keep babblers who were ignorant of mathematics from pronouncing judgment against Copernicus by reason of scripture basely twisted to suit their purpose.

Following the preface of his *De revolutionibus*, Copernicus moved on to a recitation of fundamental principles. Book 1, section 1 was titled "That the Universe is Spherical." There followed sections 2 and 3: "That the Earth is also Spherical" and "How Earth, with the Water on it, forms one Sphere." Then came

4. That the Motion of the Heavenly Bodies is Uniform, Circular, and Perpetual, or Composed of Circular Motions.

We now note that the motion of heavenly bodies is circular. Rotation is natural to a sphere and by that very act is its shape expressed. . . .

Because there are a multitude of spheres, many motions occur. Most evident to sense is the diurnal rotation . . . marking day and night. By this motion the whole Universe, save Earth alone, is thought to glide from East to West. . . .

But these bodies [the Sun, the Moon, and the five planets] exhibit various differences in their motions. First, their axes are not that of the diurnal rotation, but of the Zodiac, which is oblique thereto. Secondly, they do not move uniformly even in their own orbits; for are not Sun and Moon found now slower, now swifter in their courses? Further, at times the five planets become stationary at one point and another and even go backward [retrograde motion]. . . . Furthermore, sometimes they approach Earth, being then in Perigee, while at other times receding they are in Apogee.

Nevertheless, despite these irregularities, we must conclude that the motions of these bodies are ever circular or compounded of circles. . . .

It is then generally agreed that the motions of Sun, Moon, and Planets do but seem irregular either by reason of the divers directions of their axes of revolution, or else by reason that Earth is not the center of the circles in which they revolve, so that to us on Earth the displacements of these bodies seem greater when they are near than when they are more remote . . . Thus equal [angular] motions of a sphere viewed from different distances will seem to cover different distances in equal times. (*De revolutionibus*, I4)

A similar passage might be found in Ptolemy, the phenomena explained with eccentrics and epicycles.

In the next section, Copernicus departed from Ptolemy:

5. Whether Circular Motion belongs to the Earth; and concerning its position.

Since it has been shown that Earth is spherical, we now consider whether her motion is conformable to her shape and her position in the Universe. . . . Now authorities agree that Earth holds firm her place at the center of the Universe, and they regard the contrary as unthinkable, nay as absurd. Yet if we examine more closely it will be seen that this question is not so settled, and needs wider consideration. (*De revolutionibus*, I5)

The possibility of the Earth's motion should at least be admitted, Copernicus argued. It was, after all, the view of Heraclides and also of Ecphantus the Pythagorean. And if the Earth's possible motion is admitted, then its position should be, too. And that would enable one to adduce a reasonable cause for the apparent irregular motions of the planets:

For grant that Earth is not at the exact center but at a distance from it . . . Then calculate the consequent variation in their [the planets'] seeming motions, assuming these [motions] to be really uniform and about some center other than the Earth's. One may then perhaps adduce a reasonable cause for the irregularity of these variable motions. And indeed since the Planets are seen at varying distances from the Earth, the center of Earth is surely not the center of their circles. Nor is it certain whether the Planets move toward and away from Earth, or Earth toward and away from them. It is therefore justifiable to hold that the Earth has another motion in addition to the diurnal rotation. That the Earth, besides rotating, wanders with several motions and is indeed a Planet, is a view attributed to Philolaus the Pythagorean, no mean mathematician, and one whom Plato is said to have sought out in Italy. (*De revolutionibus*, I5)

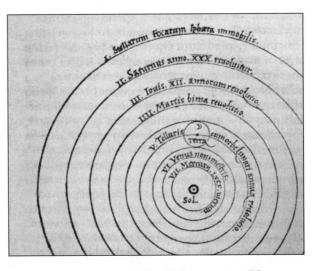

Figure 14.2: Copernicus's Heliocentric Planetary Spheres, *De revolutionibus*, 1543. The Sun (Sol) is in the center, surrounded by the spheres of Mercury, Venus, Earth, Mars, Jupiter, Saturn, and the fixed stars. Our Moon's sphere carries it around the Earth (Terra). Image copyright History of Science Collections, University of Oklahoma Libraries.

Copernicus's hypothesis did away, at least in principle, with any need for epicycles. Seemingly irregular planetary motions were instead attributed to the motion of the Earth around the Sun. In practice, though, Copernicus would be forced to use epicycles in order to save the phenomena. Indeed, he used a double epicycle for the Moon. However physically implausible the epicycle might be, it was not a violation of uniform circular motion. The equant was, and it was rejected by Copernicus for that reason.

Several phenomena are more naturally explained in Copernicus's system than in Ptolemy's. One is that Venus and Mercury are always observed in approximately the same direction as the Sun (Venus is never more than 47 degrees from the Sun, Mercury never more than 20 degrees). This phenomenon occurs inevitably in the Copernican system. In the Ptolemaic system, however, starting positions and orbital speeds must be carefully adjusted to keep the inferior planets within limited angular separations from the Sun.

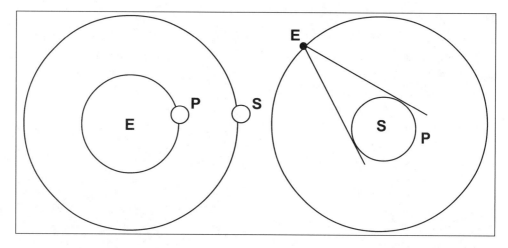

Figure 14.3: Orbit of an Inferior Planet in the Ptolemaic and Copernican Systems.
S is the Sun, *E* is the Earth, and *P* is an inferior planet (whose orbit around the Sun lies within the Earth's orbit around the Sun: either Mercury or Venus).

In the Ptolemaic model on the left, the angular speeds around the Earth of the centers of the solar and planetary epicycles must be nearly matched to keep the planet and the Sun in approximately the same angular direction as seen from the Earth. The inferior planets are never observed at opposition and must not be allowed to appear at opposition in the model.

In the Copernican model on the right, the inferior planet, *P*, as viewed from the Earth, *E*, is always seen at a small angle from the Sun, *S* (always within the angle defined by the two straight lines drawn from the Earth and encompassing the planet's orbit around the Sun). The inferior planets can never be at opposition (on the opposite side of the Earth from the Sun). Nothing further is required of the model builder to obtain this result; it is a natural, inherent, and inevitable consequence of the model.

Another phenomenon explained as a natural and inevitable consequence of Copernicus's system is that Mars, Jupiter, and Saturn are farthest from the Earth at conjunction and closest at opposition. In the Ptolemaic system, however, speeds of epicycle centers carried around by deferents, and also angular speeds of rotations of the epicycles, all had to be carefully coordinated if the observed phenomenon was to be accurately reproduced by the theory. Copernicus wrote:

> So we find . . . an admirable symmetry in the Universe, and a clear bond of har-
> mony in the motion and magnitude of the Spheres such as can be discovered in
> no other wise. For here we may observe why . . . Saturn, Jupiter, and Mars are
> nearer to the Earth at opposition to the Sun than when they are lost in or emerge
> from the Sun's rays [conjunction]. Particularly Mars, when he shines all night
> [at opposition], appears to rival Jupiter in magnitude [being closer to the Earth,
> appears brighter], being only distinguishable by his ruddy color; otherwise he
> is scarce equal to a star of the second magnitude, and can be recognized only
> when his movements are carefully followed. All these phenomena proceed from
> the same cause, namely Earth's motion. (*De revolutionibus*, I10)

The phenomena followed naturally and inevitably from the Earth's motion, leading Copernicus to assert: "These indications prove that their [the planets'] center pertains rather to the Sun than to the Earth" (*De revolutionibus*, I10). He overstated his case. The phenomena were also reproduced, albeit with consid-

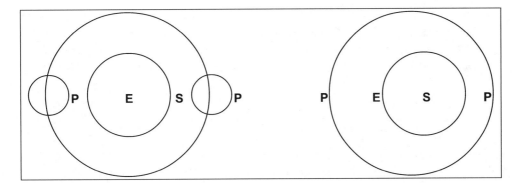

Figure 14.4: Distances from the Earth of a Superior Planet at Conjunction and Opposition in the Ptolemaic and Copernican Systems. *E* is Earth, *S* is the Sun, and *P* is a superior planet, whose orbit around the Sun lies outside Earth's orbit around the Sun.

In the Ptolemaic model on the left, the speeds of the planet's and the Sun's deferents and epicycles must all be coordinated closely to ensure that the superior planet is closer to the Earth at opposition (in the opposite direction from the Sun) and farther at conjunction (in the same direction as the Sun).

In the Copernican model on the right, the planet is naturally and inevitably closer to the Earth at opposition and farther from the Earth at conjunction.

erably more difficulty, in Ptolemy's theory. Both theories saved the phenomena. Copernicus's theory did so in a more aesthetically-satisfying manner. Therefore Copernicus's theory might be favored over its rival, everything else being equal. But the observation that Mars, Jupiter, and Saturn are farthest from the Earth at conjunction and closest at opposition is not proof that the Earth moves.

Copernicus even found admirable symmetry and harmony in the decrease of the planets' orbital sizes and times of revolution around the Sun from the outer sphere of the stars inward to the Sun:

> Given the above view—and there is none more reasonable—that the periodic times are proportional to the sizes of the Spheres . . . the order of the Spheres, beginning from the most distant is as follows. Most distant of all is the sphere of the Fixed stars, containing all things, and being therefore itself immovable. . . . Next is the planet Saturn, revolving in 30 years. Next comes Jupiter, moving in a 12-year circuit; then Mars, who goes round in 2 years. The fourth place is held by the annual revolution in which the Earth is contained, together with the Sphere of the Moon as on an epicycle. Venus, whose period is 9 months, is in the fifth place, and sixth is Mercury, who goes round in the space of 80 days. (*De revolutionibus*, I10)

The correlation between a planet's distance from the Sun and its time of revolution around the Sun was pleasing to Copernicus; it evoked in him a sense of beauty.

After showing that several planetary appearances followed from the movement of the Earth, Copernicus continued:

> But that there are no such appearances among the fixed stars argues that they are at an immense height away, which makes the circle of annual movement or its image disappear from before our eyes since every visible thing has a certain distance beyond which it is no longer seen, as is shown in optics. For the brilliance of their lights shows that there is a very great distance between Saturn

the highest of the planets and the sphere of the fixed stars. It is by this mark in particular that they are distinguished from the planets, as it is proper to have the greatest difference between the moved and the unmoved. How exceedingly fine is the godlike work of the Best and Greatest Artist! (*De revolutionibus,* I10)

The missing appearance among the fixed stars is called stellar parallax. Because the Earth circles the Sun, an observer on the moving Earth observes a star from different places over the course of a year and consequently would see the star at different angles, except that the great distance of the star makes changes in the angle of direction too small to detect. Had Copernicus been concerned with the absence of parallax as a potential refutation of his heliocentric theory, we would applaud here a great spin doctor turning an argument on its head. More likely, Copernicus was oblivious to the negative implication of the absence of a detectable stellar parallax and genuinely enthused over its positive implication. How proper, fine, and godlike it was of the best and greatest artist to have arranged to have the greatest difference in distance between the moved Earth and other planets, and the unmoved firmament of the stars.

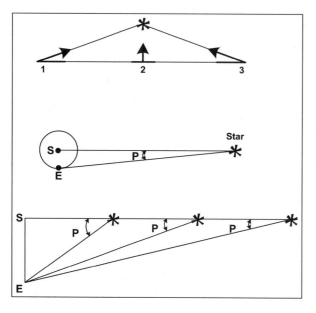

Figure 14.5: **Parallax.** top: The difference in apparent direction of an object seen from different places. As an observer moves from 1 to 2 to 3, the angular direction of the object from the observer changes. First it is seen to the right; then straight up; then back to the left.

Stellar Parallax. center: The angle subtended by the radius of the Earth's orbit at its distance from the star. S is the Sun, E the Earth, and P the parallax angle. In technical terminology, the parallax of an object is the angle that would be subtended by one astronomical unit (the mean distance of the Earth from the Sun) at the distance of the object from the Sun.

Distance. bottom: Note that the parallax angle decreases with increasing distance of the star from the Earth.

Needing to explain away the absence of any observed stellar parallax, Copernicans would argue that the great distances of the stars made any parallax angle too small to detect. Not until the 1830s, with instruments vastly better than the best in Copernicus's time, was stellar parallax first measured, by three different astronomers. The Scottish astronomer Thomas Henderson was the first to begin his measurements but the last to report them; Friedrich Struve at the Pulkovo Observatory, near St. Petersburg, was the first to publish positive results; and Friedrich Bessell, at the Königsberg Observatory, made the most convincing measurements. He measured a parallax of about a third of a second (60 seconds in one minute and 60 minutes in one degree) for the star 61 Cygni. The independent and practically simultaneous reports of the measurement of stellar parallax left little doubt that, after almost 300 years of searching and several false alarms, the stellar parallax predicted from Copernicus's heliocentric theory finally had been detected.

Interrelationships between Human Values and Science

Scientists have made a human value judgment that a theory that automatically reproduces observed phenomena is superior to a theory in which speeds, sizes, distances, and other parameters must be adjusted to reproduce the observed phenomena. Cosmological features should be explained as natural and inevitable consequences of theory, rather than merely arbitrary parameters revealed by observation. This philosophical preference guided Copernicus when he chose between rival theories that could not be separated by observations.

The same philosophical principle provided guidance at the end of the twentieth century regarding the theory of the inflationary universe. Had the initial density of the universe differed from the "critical value" by as little as 1 part in 10 to the 60th power, all matter—including living beings—would long ago have been crushed beyond recognition in a big crunch or torn apart in the expansion of a big chill. The standard big bang theory does not automatically produce the critical density. In the inflationary theory, however, a brief burst of expansion drives the density, no matter what its initial value, almost exactly to the critical value. Hence this theory is preferred.

This particular philosophical principle may be underdeveloped, or even missing, in ancient Greek thought. Greek thinkers, however, were neither blind nor deaf to linkages between human values and science. They realized that a sense of beauty could be evoked in a scientific theory. Plato regarded heaven itself and the bodies it contained as framed by the heavenly architect with the utmost beauty of which such works were susceptible. Ptolemy contemplated beautiful mathematical theories, which lifted him from Earth and placed him side by side with Zeus, where he took his fill of ambrosia, the food of the gods. Copernicus found admirable symmetry and harmony in this most beautiful temple, the universe.

Modern scientists, too, have perceived beauty in theories. Hoping to explain all the laws of physics and all the forces of nature in a single equation, cosmologists at the end of the twentieth century toyed with string theory, in which the basic constituents of the universe are tiny wriggling strings rather than particles. There was no experimental evidence for string theory, nor any prediction of observations against which to test the theory. Nonetheless, string theory persisted, largely because its mathematics was too beautiful to die.

The twentieth-century philosopher Bertrand Russell noted: "Mathematics, rightly viewed, possesses not only truth but supreme beauty—a beauty cold and austere, like that of sculpture, without appeal to any part of our weaker nature, sublimely pure, and capable of a stern perfection such as only the greatest art can show" (Russell, *Philosophical Essays,* no. 4, 73).

Analogy to sculpture, and to shape, is not out of place in the context of Greek geometrical astronomy—infatuated, as it was, with the circle. According to the twentieth-century sculptor Henry Moore: "There are universal shapes to which everybody is subconsciously conditioned, and to which they can respond if their conscious control does not shut them off" (Moore, *Henry Moore Writings and Conversations,* 195).

Modern scientists share an aesthetic sensitivity and appreciation with ancient Greek astronomers and with Copernicus. Continuation of a cultural value over centuries, even millennia, and over different civilizations is remarkable. No less remarkable would be genetic wiring of the human brain, shaping our requirements for an aesthetically satisfying understanding of nature. Such speculation is premature but does suggest further study of the nature of ancient and modern science, especially similarities between their human values and ways of thinking.

The Hypothetico-Deductive Method

Scientific theories can be tested by the *hypothetico-deductive method*. First, a hypothesis is postulated: that the Earth orbits around the Sun. Then, a prediction is deduced from the hypothesis: the motion of the Earth will be revealed in a stellar parallax. Then the test: is a stellar parallax found?

In this historical case, the answer was no; and the logic of the method dictated that the original hypothesis was thus refuted: the Copernican theory was false. Or an auxiliary hypothesis had been overlooked: that the stars are very distant and thus a parallax angle may actually exist but is too small to detect.

Recourse to an auxiliary hypothesis in an attempt to rescue a cherished major hypothesis after it has been observationally refuted constitutes an ex post facto argument. In science, such arguments made after a fact (in this case, no stellar parallax) becomes known or a theory is modified to bring it into agreement with new facts, are less psychologically compelling than are predictions of previously unknown phenomena, even if the strict logic of the two situations is equally compelling.

The hypothetico-deductive method has the potential—logically, at least in principle—to refute a hypothesis, if all potentially extenuating auxiliary assumptions are included in the analysis.

But the hypothetico-deductive method cannot be used to prove a hypothesis. When a theory is tested using several predictions, and all the predicted phenomena are found, one reasonably gains confidence in the theory. But observations do not (can never) constitute absolute proof. When observations match prediction, a theory commands greater confidence, but a theory is never (can never be) incontrovertibly proven.

In attempting to prove the Copernican theory, consider the following logic:

hypothesis: the Copernican theory
prediction: if this theory, then Mercury and Venus will always be near the Sun
observation: Mercury and Venus are always observed near the Sun
conclusion: therefore the Copernican theory

But is this conclusion warranted? Aren't Mercury and Venus always seen near the Sun in the Ptolemaic model too? We have here a case of the fallacy of affirming the consequent: If A, then B; B; therefore A.

Remember the medieval nominalist argument: that we cannot insist upon the truth of any particular working hypothesis because God could have made the world in some different manner that nonetheless has the same set of observational consequences.

A God capable of creating the world in a different manner isn't even necessary: just some ingenious person to imagine such a possibility. Ptolemy kept Mercury and Venus always near the Sun, but not as an inevitable consequence of his geometric model. The desired naturalness and inevitability can, however, be achieved in an Earth-centered system. Were the Sun to orbit around a motionless Earth in the center while Mercury and Venus orbited around the Sun, Mercury and Venus would always be seen near the Sun. Indeed, just such a model would be proposed as a rival to the Copernican theory.

An unauthorized preface added to Copernicus's *De revolutionibus* presented Copernicus's theory as a mathematical fiction rather than the true account of the beauty and harmony of the universe that Copernicus intended. The spurious preface read, in part:

For it is the job of the astronomer to use painstaking and skilled observation in gathering together the history of the celestial movements, and then—since he cannot by any line of reasoning reach the true causes of these movements—to think up or construct whatever causes or hypotheses he pleases such that, by the assumption of these causes, those same movements can be calculated from the principles of geometry for the past and for the future too. . . . [I]t is not necessary that these hypotheses should be true, or even probable; but it is enough if they provide a calculus which fits the observations . . . [L]et no one expect anything, in the way of certainty from astronomy, since astronomy can offer us nothing certain, lest if anyone take as true that which has been constructed for another use, he go away from this discipline a bigger fool than when he came to it. Farewell. (*De revolutionibus*, introduction)

Andrew Osiander, a Lutheran clergyman who tended Copernicus's book through the printing process, added this preface without Copernicus's knowledge. Copernicus saw the book only as he was dying, and was unable to make any changes.

Copernicus was a mathematical realist, not an instrumentalist or a nominalist. He believed passionately that the admirable symmetry and harmony in motions and magnitudes contained within his astronomical model was proof of its reality. Observed phenomena that were ad hoc arbitrary arrangements in Ptolemy's geocentric model occurred automatically and inevitably in Copernicus's heliocentric model. This element of completeness gave Copernicus confidence that he had discovered the real order of the cosmos, not merely a mathematical fiction.

Confidence and passion energized Copernicus and led to the scientific breakthrough only hinted at in the earlier hypothetical musings of nominalists such as Buridan and Oresme. Copernicus the realist exalted:

> **Copernicus's Grave**
>
> Copernicus was buried in a tomb under the floor of the Roman Catholic cathedral in Frombork, 180 miles north of Warsaw. In the summer of 2005, after a year of searching, Polish archeologists located what is very likely Copernicus's grave. Police forensic experts determined that the skull belonged to a man who died at about age 70, and a computer reconstruction of the face matched portraits of Copernicus with a broken nose and a scar above his left eye. Scientists will try to find relatives of Copernicus and do DNA identification.

In the middle of all sits Sun enthroned, in this most beautiful temple could we place this luminary in any better position from which he can illuminate the whole at once? He is rightly called the Lamp, the Mind, the Ruler of the Universe . . . So the Sun sits upon a royal throne ruling his children the planets which circle round him. The Earth has the Moon at her service. . . . [T]he Earth conceives by the Sun, and becomes pregnant with an annual rebirth. (*De revolutionibus*, 110)

For all that Copernicus accomplished, his astronomy remained the astronomy of the Greeks and of Ptolemy, employing combinations of uniform circular motions to save the phenomena. A simple change in geometry, switching the Earth and the Sun, had little affect on working astronomers. Either system

> ## Mental Exercise: Greater and Lesser Progressions and Retrogressions
>
> In addition to distances at conjunction and opposition, there also followed naturally and inevitably from Copernicus's theory explanations for "why the progression and retrogression appear greater for Jupiter than Saturn, and less than for Mars, but again greater for Venus than for Mercury; and why such oscillation appears more frequently in Saturn than in Jupiter, but less frequently in Mars and Venus than in Mercury. . . ." (*De revolutionibus*, 110). Explain, with the use of diagrams, how these phenomena follow inevitably from Copernicus's heliocentric model.

could be used for calculating planetary positions. Neither system then commanded an observational advantage over the other. *De revolutionibus* was not a revolutionary book. But it was revolution-making. How switching the positions of the Sun and Earth initiated a scientific revolution is the subject of the following chapter.

15

THE COPERNICAN REVOLUTION

Revolutions in science often surpass the limited changes envisioned by their initiators, much as political and social revolutions do. Such is the case with the Copernican revolution.

It began with a seemingly simple interchange of the Earth and the Sun in a geometrical model still composed of uniform circular motions. But Aristotelian physics, with objects falling to the Earth because of their natural tendency to move to the center of the universe (where the Earth was), does not work in a Copernican universe. The old physics would be destroyed and eventually replaced by a new, Newtonian physics. Other worlds, forbidden in an Aristotelian universe because their earthy nature would have caused them to move to their natural place at the center of the universe, now could be scattered throughout a Copernican universe.

The revolution would also see the demise of the finite, closed, and hierarchically ordered universe of medieval belief. It would be replaced by an indefinite or even infinite universe consisting of components and laws but lacking value concepts: perfection, harmony, meaning, and purpose. No longer was the universe specifically created for humankind, which was further downgraded by the new possibility of many intelligent species scattered throughout the universe. The Earth was now but one of many planets similar in physical composition and possibly with similar inhabitants. As scientists were changing the physical view of the universe, other thinkers would radically alter political, sociological, and religious views of humankind's place and role in the new universe.

Back on the physical side of the revolution, once the apparent motion of the stars was transferred from rotation of their sphere to rotation of the Earth, the outer sphere of the stars was obsolete. Only intellectual inertia discouraged speculations about newly possible distributions of stars throughout a perhaps infinite space and speculations about noncentral positions for the Earth and the Sun.

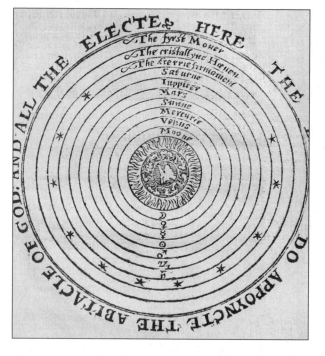

Figure 15.1: **Leonard Digges, *A prognostication everlasting*, 1556.** The ancient and medieval picture of the universe. The sublunar region consists of spheres of earth, water, air, and fire. Beyond are the spheres of the Moon, the Sun, the planets, and the stars.

Illustration from Leonard Digges, *A prognostication everlasting*. 1556. Image copyright History of Science Collections, University of Oklahoma Libraries.

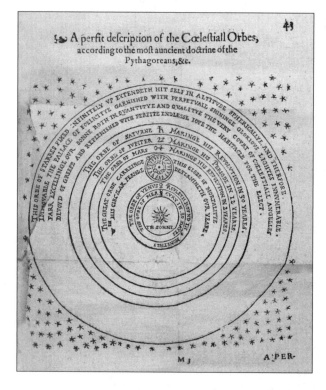

Figure 15.2: **Thomas Digges, *A Perfit Description of the Caelestial Orbes*, 1576.** This diagram was added as an appendix by Thomas Digges to his father Leonard Digges' 1576 *A prognostication everlasting*, "to the end [that] such noble English minds (as delight to reach above the baser sort of men) might not be altogether defrauded of so noble a part of Philosophe."

Illustration from Thomas Digges, *A Perfit Description of the Caelestial Orbes*. 1576. Image copyright History of Science Collections, University of Oklahoma Libraries.

The English astronomer Thomas Digges began to realize logical consequences of Copernicus's new system. His father, Leonard Digges, had published in 1556 a standard medieval diagram of the universe with the Earth in the center. In 1576 Thomas reprinted his father's book with a new appendix

Galileo, Experiments, the Leaning Tower of Pisa, and Pendulums

According to biographical notes made by a disciple, companion, and pupil of Galileo for the three years immediately preceding his death in 1642 and published a dozen years later:

> [T]o the great confusion of all the philosophers, very many conclusions of Aristotle himself about the nature of motion, which had been theretofore held as most clear and indubitable, were convicted of falseness by means of experiments and by sound demonstrations and discourses; as among others, that the velocity of moving bodies of the same composition, unequal in weight, moving through the same medium, do not attain the proportion of their weights, as Aristotle assigned to them, but rather that they move with equal velocity, proving this by repeated experiments performed from the summit of the Campanile of Pisa, in the presence of all other teachers and philosophers and of all the students. (Holton, *Physics*, 51)

From these words grew the legend that Galileo dropped simultaneously from the Leaning Tower of Pisa iron cannon balls of different weights, and that they hit the ground simultaneously. The story could well be true. Galileo, however, never claimed credit for any such demonstration while he was alive—an unlikely omission had he really carried out the purported experiment.

The story serves to substantiate the myth that all scientific theories originate in experiment or observation. An alternative view to Galileo *empiricist* is Galileo *theorist*. He used logic rather than, or at least in addition to, experiment to examine and understand physical laws. In a dialog in one of his books attacking and demolishing Aristotelian physics, Galileo had his wise philosopher argue:

> But without other experiences, by a short and conclusive demonstration, we can prove clearly that it is not true that a heavier moveable [body] is moved more swiftly than another, less heavy . . . then if we had two moveables whose natural speeds were unequal, it is evident that were we to connect the slower to the faster, the latter would be partly retarded by the slower, and this would be partly speeded up by the faster . . . But if this is so, and if it is also true that a large stone is moved with eight degrees of speed, for example, and a smaller one with four, then joining both together, their composite will be moved with a speed less than eight degrees. But the two stones joined together make a larger stone than the first one which was moved with eight degrees of speed; therefore this greater stone is moved less swiftly than the lesser one. But this is contrary to your assumption. So you see how, from the supposition that the heavier body is moved more swiftly than the less heavy, I conclude that the heavier moves less swiftly. (Holton, *Physics*, 81)

Galileo's thought experiment is more compelling than an attempt in the 1980s to repeat the supposed historical experiment. Unfortunately for pedagogues in search of simple classroom illustrations, the lighter ball dropped from the Leaning Tower hit the ground first! (The experimenter believed that he had released the balls simultaneously, but perhaps fatigue in his hand holding the heavier ball caused that hand to let go more slowly?)

Another legend has Galileo's study of pendulums stimulated by observing a hanging incense burner swinging back and forth in the Pisa Cathedral when he was a student, in 1583. Supposedly, Galileo noticed that each swing took the same time, whether the arc of the swing was large or small. This legend might also be true, although Galileo didn't develop his theory of pendulums until 1638, and the huge bronze incense burner in the central aisle now pointed to as Galileo's inspiration was installed after a fire severely damaged the interior of the cathedral in 1596.

in which he, Thomas, described the great orb of the heavens as "garnished with lights innumerable and reaching up in *Sphaericall altitude* without end" (Hetherington, *Encyclopedia of Cosmology,* 93–94, 176–77). His diagram depicts stars scattered at varying distances beyond the former boundary of the sphere of the stars and notes that the stars extend infinitely upwards. The diagram also illustrates continued obliviousness to another logical consequence of the Copernican model: Thomas accorded to the Sun a unique place at the center of the universe, even though the concept of a center loses all meaning in an infinite universe.

The idea of the Moon as a planet similar to the Earth would be dramatically emphasized by Galileo Galilei's telescopic discoveries beginning in 1610. He was born in 1564, the year of Shakespeare's birth and Michelangelo's death. His father wanted him to study medicine, which Galileo did at the University of Pisa. But he preferred mathematics and studied it with private tutors in Pisa and then at home in Florence. Soon he was giving private lessons and then was appointed to the mathematics chair at the University of Pisa when he was only 26 years old. Obstinate and argumentative, Galileo was unpopular with the other professors. His experiment involving cannon balls dropped from the Leaning Tower of Pisa—if it really occurred—was a public challenge to Aristotelian philosophers.

In 1592 Galileo left Tuscany to take up the chair of mathematics at the University of Padua, in the Republic of Venice. He would stay there 18 years. In 1609 rumor reached Galileo from Holland of a device using pieces of curved glass to make distant objects on the Earth appear near. He quickly constructed his own telescope and turned it to the heavens.

One of the first objects he viewed was the Moon. Large dark spots had been seen by many observers before Galileo, but his far more detailed telescopic observations more emphatically demanded the revolutionary conclusion that the Moon was not a smooth sphere, as Aristotelians had maintained, but was uneven and rough, like the Earth. Discernible features of the Moon—the so-called man in the moon—might be attributed by followers of Aristotle to vapors and mists terrestrial in origin or to lunar mountains and craters under a smooth, transparent covering material. But such ex post facto arguments were no match for Galileo's new telescopic observations.

In his *Sidereus nuncius (Starry Messenger)* of 1610, Galileo reported his observations of numerous small spots on the Moon. The boundary dividing the dark and light parts of the Moon a few days after new moon was not a uniform oval line (as it would have been for a smooth sphere) but was uneven and wavy. Within the illuminated region were a few bright spots. Galileo wrote: "The surface of the Moon is not smooth, uniform, and precisely spherical as a great number of philosophers believe it to be, but is uneven, rough, and full of cavities and prominences, being not unlike the face of the Earth, relieved by chains of mountains and deep valleys" (Drake, *Discoveries,* 31). The analogy to mountains on the Earth was further strengthened by Galileo's observation that the blackened parts of lunar spots were always directed

Figure 15.3: Frontispiece, Galileo Galilei, *Sidereus nuncius,* **1610. ASTRONOMICAL / MESSENGER /** great and admirable sights / displayed to the gaze of everyone, but especially / *PHILOSOPHERS and ASTRONOMERS*, observed by / **GALILEO GALILEO / Florentine patrician /** University of Padua public mathematician / **with a spyglass /** devised by him, about the face of the Moon, countless fixed stars, / THE MILKY WAY, NEBULOUS STARS, / but especially / **FOUR PLANETS /** around the star JUPITER at unequal intervals, periods / with wonderful swiftness; by anyone / until this day unknown, / the author detected and decided / to name first / **MEDICEAN STARS.**

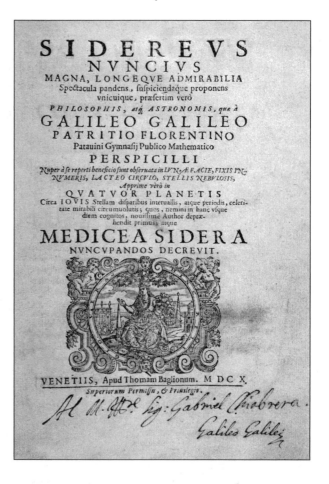

toward the Sun, while the opposite side was bright, like a shining mountain peak.

Near new moon, the portion of the Moon shadowed from direct sunlight was slightly brightened. Ruling out an inherent and natural light of the Moon or reflected light from Venus, Galileo argued that the source of light was sunlight reflected from the Earth: *earthshine.* Thus the Earth was not devoid of light, but was similar to bodies such as the Moon and Venus.

Galileo's observations of satellites of Jupiter furnished another similarity between the Earth and a planet: both had moons circling them. Thus there was more than one center of motion, as the Copernican system asserted and the Ptolemaic system had denied. But much as Galileo wanted proof of the Copernican system, Tycho Brahe's planetary system (discussed in the following chapter), a compromise adopted by many influential Jesuit astronomers, also had more than one center of motion. Theories may be refuted by observation but not proved, because some other theory with the same observational consequences is always possible. Still, Galileo's observations strengthened the Copernican case.

Galileo's drawing of the Moon, 1610

Galileo's drawing of the Moon includes a large round crater on the terminator (the dividing line between lit and unlit hemispheres), a nearby field of smaller craters on the bright side of the terminator, and three bright arms crossing the terminator into the dark hemisphere.

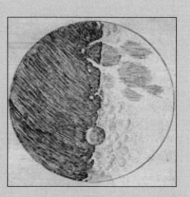

These features are not found in modern photographs of the Moon. Perhaps Galileo was obliged to depict in his small engraving the crater much larger than it actually is so he could show the contrasting illumination of the rim of the crater at first and last quarter. Or perhaps Galileo's drawing was not entirely an artifact of observation free of mental assistance. The size of the enormous round crater near the terminator may be commensurate with the psychological impact on Galileo's thinking of Albategnius, an actual and considerably smaller crater.

Early telescopes rarely were powerful enough for naïve, unbiased observers to recognize what they were seeing. The earliest extant record of an astronomical observation made with a telescope is the Englishman John Harriot's lunar drawing of July 26, 1609. The shading is haphazard, with no suggestion of craters and mountains, and the terminator curves in an impossible manner. Correspondence between Harriot and his friends reveals their initial confusion; they had no idea that they were seeing shadows cast by craters and mountains.

The ability to see often was dependent less on visual acuity and more on the enhancing effect of theoretical preconception. By the summer of 1610 Harriot and his friends were discussing Galileo's reported observations, and Harriot could have seen a copy of Galileo's Sidereus *nuncius*, or *Starry Messenger*, by early July. Harriot's July 1610 drawing of the Moon contains several idiosyncratic features (an enormous crater, three jagged protrusions, and a field of craters), all of which also are found in Galileo's drawing, and none of which are found in modern photographs of the Moon. Belief was transformed into sight; believing had become seeing. Image copyright History of Science Collections, University of Oklahoma Libraries.

On January 7 in the year 1610, Galileo noticed three objects, small but very bright, near Jupiter—two on the eastern side of the planet, and one to the west. He believed them to be stars, but his curiosity was aroused by the fact that they lay in a straight line parallel to the ecliptic. The next night, all three were to the west of Jupiter. Galileo wrote: "Hence it was with great interest that I awaited the next night. But I was disappointed in my hopes, for the sky was then covered with clouds everywhere" (Drake, *Discoveries*, 52). On January 10, two of the objects were to the east, and one was presumably hidden behind Jupiter. On the next night, there were two again to the east, but one much brighter than the other and farther from Jupiter than on the previous night. Galileo wrote: "I had now decided beyond all question that there existed in the heavens three stars wandering about Jupiter as do Venus and Mercury about the Sun" (Drake, *Discoveries*, 53). On January 13, Galileo saw four stars for the first time. He concluded:

> Above all, since they sometimes follow and sometimes precede Jupiter by the
> same intervals, and they remain within very limited distances either to east

or west of Jupiter, accompanying that planet in both its retrograde and direct movements in a constant manner, no one can doubt that they complete their revolutions about Jupiter . . .

Here we have a fine and elegant argument for quieting the doubts of those who, while accepting with tranquil mind the revolutions of the planets about the Sun in the Copernican system, are mightily disturbed to have the Moon alone revolve about the Earth and accompany it in an annual rotation about the Sun. Some have believed that this structure of the universe should be rejected as impossible. But now we have not just one planet rotating about another while both run through a great orbit around the Sun; our own eyes show us four stars which wander around Jupiter as does the Moon around the Earth, while all together trace out a grand revolution about the Sun in the space of twelve years. (Drake, *Discoveries*, 56–57)

Figure 15.5: Galileo's Discovery of Four Satellites of Jupiter. A Page from Galileo, *Sidereus nuncius*, 1610.

Galileo regarded as his most important discovery four planets (satellites of Jupiter) never seen before. Beginning in January 1610, Galileo observed Jupiter and discovered three, and then four, little stars positioned near the planet, arranged along a straight line, parallel to the ecliptic. He reported their disposition among themselves and with respect to Jupiter. In his *Sidereus nuncius*, Galileo reported:

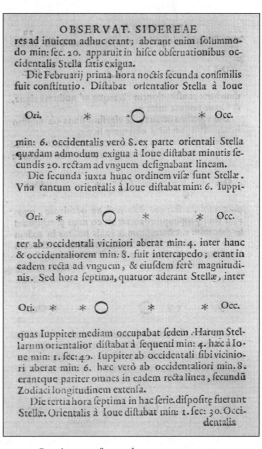

On February first, the second hour of the night, the formation was similar. The eastern star 6 minutes from Jupiter, the western one 8. To the east a very small star 20 seconds distance from Jupiter, in a straight line.

On February second, a star to the east 6 minutes from Jupiter. Jupiter 4 minutes from the nearer one to the west; and 8 minutes between the two western stars. In a straight line and of nearly equal magnitude. At the seventh hour there were four stars, Jupiter occupying the middle. The easternmost 4 minutes from the next, and that 1 minute, 40 second from Jupiter. Jupiter 6 minutes from the closest western one, and that one 8 minutes from the westernmost one. They were on the same straight line, an extension in longitude of the zodiac.

Source: Translation after Albert van Helden, *Sidereus Nuncius or The Sidereal Messenger. Galileo Galilei. Translated with introduction, conclusion, and notes by Albert van Helden* (Chicago: University of Chicago Press, 1989), p. 74.

Image copyright History of Science Collections, University of Oklahoma Libraries.

Galileo named the four objects never before seen the Medicean Stars, after Cosimo II de Medici, the Grand Duke of Tuscany. In his introduction to the *Starry Messenger*, Galileo noted that marble and bronze statues, columns, and pyramids are constructed to pass down for the memory of posterity names deserving immortality. All, however, perish in the end. The fame of Jupiter, Mars, Mercury, and Hercules, on the other hand, enjoy with the stars eternal life. Now, just as the immortal graces of Cosimo's soul had begun to shine forth on Earth, bright stars offered themselves like tongues to speak of and celebrate his most excellent virtues for all time. These words would not have seemed overly obsequious to Galileo's contemporaries living in a hierarchically ordered society. Then, kings and dukes subsidized economically unproductive occupations such as painting, music, literature, and science. Now, democracies need to find replacements for royal patronage.

That spring, at Easter, Galileo traveled to Florence to demonstrate the use of the telescope and personally show the new stars to Cosimo. Cosimo responded as Galileo had hoped, inviting Galileo to take up residence in Florence that July as mathematician and philosopher to the grand duke, and also as chief mathematician of the University of Pisa, without obligation to teach.

That fall or winter, Galileo focused an improved telescope on Venus, which had moved on from its summer conjunction with the Sun and was now visible in the evening sky. Were Venus a planet like the Earth, rather than a wandering self-illuminated star, and shining with light borrowed from the Sun, it should show phases, as does the Moon. A former pupil wrote to Galileo: "Since (as I believe) the opinion of Copernicus that Venus revolves about the Sun is correct, it is clear that she would necessarily be seen by us sometimes horned and sometimes not . . . Now I would like to know from you if you, with your wonderful spyglasses, have noticed such an appearance, which, without doubt, will be a sure means of convincing any obstinate mind" (van Helden, *Sidereus Nuncius*, 106).

To ensure credit to himself for just such a discovery, Galileo immediately wrote to the Tuscan ambassador in Prague that he had observed a phenomenon arguing strongly for the Copernican system. Perhaps Galileo only began to study Venus after being alerted to the possibility of phases: he stole credit for the idea. Or perhaps Galileo had been observing Venus for months, as he later claimed: all the credit is rightfully his, for the idea as well as for the actual observation.

Venus now was at half-moon, not yet crescent with horns. Until then, Galileo dared not specify the phenomenon. Instead, he presented the ambassador with an anagram. Its solution, as Galileo later revealed, was "the mother of love emulates the figures of Cynthia" (van Helden, *Sidereus Nuncius*, 107). (The mother of love is Venus, Cynthia is the Moon, and the figures of Cynthia are the phases of the Moon.)

Venus duly passed from half-moon to crescent phase, and Galileo replied to his former pupil:

Know therefore, that about 3 months ago I began to observe Venus with the instrument, and I saw her in a round shape and very small. Day by day she increased in size and maintained that round shape until finally, attaining a very great distance from the Sun, the roundness of her eastern part began to diminish, and in a few days she was reduced to a semicircle. She maintained this shape for many days, all the while, however, growing in size. At present she is becoming sickle-shaped, and as long as she is observed in the evening her little horns will continue to become thinner, until she vanishes. But when she then reappears in the morning, she will appear with very thin horns again turned away from the Sun, and will grow to a semicircle at her greatest digression [greatest distance from the Sun]. She will then remain semicircular for several days, although diminishing in size, after which in a few days she will progress to a full circle. Then for many months she will appear, both in the morning and then in the evening, completely circular but very small. (van Helden, *Sidereus Nuncius*, 107–8)

The increase in apparent size was the result of Venus coming nearer the Earth. Galileo's observations were consistent with the Copernican system but not with the Ptolemaic system. Venus changed shape precisely as does the Moon.

Galileo concluded, in another letter: "These things leave no room for doubt about the orbit of Venus. With absolute necessity we shall conclude, in agreement with the theories of the Pythagoreans and of Copernicus, that Venus revolves about the Sun just as do all the other planets. Hence it is not necessary to wait for transits and occultations of Venus to make certain of so obvious a conclusion" (Drake, *Discoveries*, 93–94). A transit of Venus occurs when the planet passes across the face of the Sun; an occultation occurs when a body passes behind the Sun. The tiny speck of Venus on the face of the Sun would not be observed until 1639.

Another telescopic observation with revolutionary implications was of sunspots. Several astronomers, including Galileo, independently observed them in 1610 and 1611. The Sun was found to be blemished, as were the Earth and the Moon. Not only did this observation contradict Aristotelian physics, but it would open the door to extending physical principles applicable to terrestrial phenomena also to celestial phenomena, as Isaac Newton would famously do later in the century.

In an attempt to save the perfection of the supposedly incorruptible and unalterable Sun, a devout Aristotelian asserted that the observed phenomena were not blemishes on the face of the Sun but instead satellites of the Sun. In response, Galileo began a systematic study of sunspots in 1612. (Direct observation of the Sun is dangerous, and Galileo eventually went blind, though he mainly observed a projected image. This observational technique was discovered by one of his pupils.)

First, Galileo established that the spots were not "mere appearances or illusions of the eye or of the lenses of the telescope." He wrote that the spots do not "remain stationary on the body of the Sun, but appear to move in relation

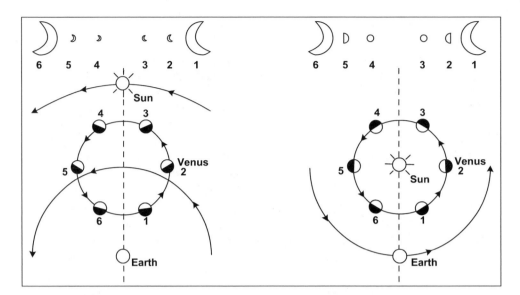

Figure 15.6: Phases of Venus in Ptolemaic and Copernican Models. In both the Ptolemaic geocentric model on the left and the Copernican heliocentric model on the right, Venus appears much larger when it is closest to the Earth, between *6* and *1*, than when it is farthest from the Earth, between *3* and *4*.

In the Copernican model, Venus, shining with light borrowed from the Sun, exhibits a full range of phases, from new (between *6* and *1*) through crescent to full (between *3* and *4*). In the Ptolemaic model, however, Venus does not display a full phase to observers on the Earth.
Source: After Albert van Helden, *Sidereus Nuncius or The Sidereal Messenger. Galileo Galilei. Translated with introduction, conclusion, and notes by Albert van Helden* (Chicago: University of Chicago Press, 1989), p. 108.

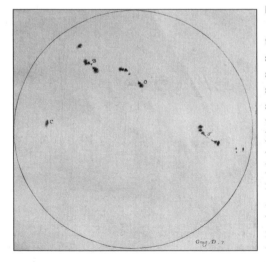

Figure 15.7: Sunspot Drawing by Galileo, summer 1612. Sunspots are irregularly shaped dark areas on the surface of the Sun, often large enough (tens of thousands of miles in diameter) to be visible without a telescope. The Greek philosopher Anaxagoras may have seen a sunspot in 467 B.C., and Chinese records of sunspots date from 28 B.C. According to Aristotle, however, the heavens were perfect and unchanging. Hence, a large spot persisting for eight days in 807 was thought to be Mercury passing in front of the Sun. The number of sunspots has a 22-year cycle from minimum to maximum activity and back to minimum activity. From A.D. 1645 to A.D. 1715 sunspots were very rare; fewer were seen during the entire period than are now seen in a single year. Climate changes seem to be linked to solar activity. Image copyright History of Science Collections, University of Oklahoma Libraries.

to it with regular motions." Indeed, the spots described "lines on the face of the Sun similar to those along which Venus and Mercury proceed when those planets come between the Sun and our eyes" (Drake, *Discoveries*, 91).

Nor were the spots objects located in the Earth's atmosphere. That close, they would have displayed an easily perceptible parallax.

Nor were sunspots satellites circling the Sun. The companions of Jupiter were constant in appearance, with regular and definite periods. Sunspots, on the other hand, were generated and decayed in longer and shorter periods. Some condensed and others expanded, and some were most irregular. And their speeds across the face of the Sun varied.

From analogy with terrestrial phenomena, it clearly followed "that no material of ours better imitates the properties of these spots than terrestrial clouds" (Drake, *Discoveries*, 99). In Galileo's opinion: "[T]he solar spots are produced and dissolve upon the surface of the Sun and are contiguous to it, while the Sun, rotating upon its axis in about one lunar month, carries them along, perhaps bringing back some of those that are of longer duration than a month, but so changed in shape and pattern, that it is not easy for us to recognize them" (Drake, *Discoveries*, 102).

Alterable spots on the surface of the Sun were yet another weapon for Galileo to wield in his vicious verbal battle against Aristotelians (and, by implication, in favor of the alternative Copernican astronomy). He wrote:

> I believe that there are not a few Peripatetics [Aristotelians] . . . who go about philosophizing without any desire to learn the truth and the causes of things, for they deny these new discoveries or jest about them, saying that they are illusions. It is about time for us to jest right back at these men and say that they likewise have become invisible and inaudible. They go about defending the inalterability of the sky, a view which perhaps Aristotle himself would abandon in our age. . . . People like this, it seems to me, give us reason to suspect that they have not so much plumbed the profundity of the Peripatetic arguments as they have conserved the imperious authority of Aristotle. (Drake, *Discoveries*, 140–41)

Albert Einstein recognized in Galileo a passionate fighter against dogma based on authority.

Galileo also confronted and confounded arguments from Aristotelian physics raised against the Copernican system. In his *Dialogue Concerning the Two Chief World Systems—Ptolemaic & Copernican*, published in 1632, Galileo took

> the Copernican side in the discourse, proceeding as with a pure mathematical hypothesis and striving by every artifice to represent it as superior to supposing the Earth motionless—not, indeed, absolutely, but as against the arguments of some professed Peripatetics [Aristotelians]. These men indeed deserve not even that name [*peripatetic* means walking about], for they do not walk about; they are content to adore the shadows, philosophizing not with due circumspection but merely from having memorized a few ill-understood principles. (Drake, *Dialogue*, 6–7)

Mental Exercise: Sunspots

Galileo observed that the spaces passed by the same sunspot in equal times became always less as the spot was situated nearer the edge of the Sun. Explain why this phenomenon occurs and its potential significance in a debate over whether sunspots are contiguous to the Sun or satellites of the Sun. Can you quantify the argument? For help, visit http://solar-center. stanford.edu/sunspots/gproof.html.

Figure 15.8: Frontispiece, Portrait of Galileo, *The Assayer*, 1623. In 1618 Orazio Grassi, professor of mathematics at the Collegio Romano, gave a public lecture on comets. A manuscript copy of the lecture was sent to Galileo, and the lecture was published in 1619 as *On the Three Comets of the Year MDCXVIII. An Astronomical Disputation Presented Publicly in the Collegio Romano of the Society of Jesus by one of the Fathers of that same Society.* Galileo drafted a critique of Grassi's work, and a pupil published it as *Discourse on the Comets. By Mario Guiducci. Delivered at the Florentine Academy during his Term as Consul.* Grassi angrily replied in a book of his own, though under the pseudonym Lothario Sarsi: *The Astronomical Balance, on which the Opinions of Galileo Galilei regarding Comets are weighed, as well as those presented in the Florentine Academy by Mario Guiducci and recently published.* In 1623 Galileo's friend and supporter Cardinal Maffeo Barberini became Pope Urban VIII, and Galileo replied to Grassi's attack with *The Assayer*, dedicated to the pope, who then told Galileo that he could write about the Copernican theory as a mathematical hypothesis. Image copyright History of Science Collections, University of Oklahoma Libraries.

That Galileo was presenting merely a mathematical hypothesis, not necessarily the true system of the world, was itself a fiction. This fiction was necessary because in 1616 the Congregation of the Index had issued an edict forbidding assertion of literal truth for the motion of the Earth and the stability of the Sun. But hypothetical discussions of Copernican astronomy were permissible.

Galileo wrote in his preface to the discerning reader that he accepted the "salutary edict which, in order to obviate the dangerous tendencies of our present age, imposed a seasonable silence upon the Pythagorean opinion that the Earth moves" (Drake, *Dialogue*, 5). In answer to "those who impudently asserted that this decree had its origin not in judicious inquiry, but in passion none too well informed," Galileo would "show to foreign nations that as much is understood of this matter in Italy, and particularly in Rome [i.e., the Church], as transalpine [the Protestant Reformation was occurring north of the Alps] diligence can ever have imagined" (Drake, *Dialogue*, 5). This was just what the pope wanted: a demonstration that the Catholic Church did not interfere with the pursuit of science.

Galileo's dialogue was between Salviati, "a sublime intellect which fed no more hungrily upon any pleasure than it did upon fine meditations," and Simplicio, "a Peripatetic philosopher whose greatest obstacle in apprehending the truth seemed to be the reputation he had acquired by his interpretations of Aristotle" (Drake, *Dialogue*, 7). Sagredo, "a man of noble extraction and trenchant wit" (Drake, *Dialogue*, 7), was mainly an onlooker to the debate but also an occasional commentator. Simplicio, though neither as smart nor as well informed as Salviati, was not a simpleton. He was named after the sixth-century philosopher Simplicius, who wrote commentaries on Aristotle's *Physics* and *On the Heavens*. In Simplicio's mouth, Galileo unwisely placed arguments personally made to him by the pope.

Figure 15.9: Frontispiece, *Dialogo di Galileo Galilei*. The *Dialogo sopra i due massimi sistemi del mondo* (*Dialogue Concerning the Two Chief World Systems*) was published in Florence in 1632 under a formal license from the Inquisition. The title on the title page is *Dialogo*, followed by Galileo's name and references to his membership in the Accademia dei Lincei and his employment then under the Grand Duke Ferdinand II of Tuscany, son of Cosimo II, Galileo's earlier patron who had died in 1620. Image copyright History of Science Collections, University of Oklahoma Libraries.

Salviati: As the strongest reason of all [against motion of the Earth] is adduced that of heavy bodies, which, falling down from on high, go by a straight and vertical line to the surface of the Earth. This is considered an irrefutable argument for the Earth being motionless. For if it made the diurnal rotation, a tower from whose top a rock was let fall, being carried by the whirling of the Earth, would travel many hundreds of yards to the east in the time the rock would consume in its fall, and the rock ought to strike the Earth that distance away from the base of the tower. (Drake, *Dialogue*, 126)

Similarly, a lead ball dropped from the mast of a moving ship would drop at a distance from the mast; a projectile thrown to a great height would fall as far to the west as the Earth had carried the thrower eastward; and a canon ball shot to the west ought to range much farther than one shot to the east, adding the distance the cannon was carried by a moving Earth to the distance of one shot, and subtracting it from the other. The dialogue continued:

Salviati: Now experiment shows the shots to fall equally; therefore the cannon is motionless, and consequently the Earth is, too. Not only this, but shots to the south or north likewise confirm the stability of the Earth; for they would never hit the mark that one had aimed at, but would always slant toward the west because of the travel that would be made toward the east by the target, carried by the Earth while the ball was in the air.

Simplicio: Oh, these are excellent arguments, to which it will be impossible to find a valid answer.

Salviati: Perhaps they are new to you?

Simplicio: Yes, indeed, and now I see with how many elegant experiments nature graciously wishes to aid us in coming to the recognition of the truth . . .

Sagredo: What a shame there were no cannons in Aristotle's time! With them he would indeed have battered down ignorance, and spoken without the least hesitation concerning the universe.

Salviati: It suits me very well that these arguments are new to you, for now you will not remain of the same opinion as most Peripatetics, who believe that anyone who departs from Aristotle's doctrine must therefore have failed to understand his proofs. . . .

Tell me Simplicio Do you feel convinced that the experiment on the ship squares so well with our purpose that one may reasonably believe that whatever is seen to occur there must also take place on the terrestrial globe? . . .

Simplicio: So far, yes . . .

Salviati: Rather, I hope that you will stick to it, and firmly insist that the result on the Earth must correspond to that on the ship, so that when the latter is perceived to be prejudicial to your case you will not be tempted to change your mind . . .

Very good. Now, have you ever made this experiment of the ship?

Simplicio: I have never made it, but I certainly believe that the authorities who adduced it had carefully observed it. Besides, the cause of the difference is so exactly known that there is no room for doubt.

Salviati: You yourself are sufficient evidence that those authorities may have offered it without having performed it, for you take it as certain without having done it, and commit yourself to the good faith of their dictum. Similarly it not only may be, but must be that they did the same thing too—I mean, put faith in their predecessors, right on back without ever arriving at anyone who had performed it. For anyone who does will find that the experiment shows exactly the opposite of what is written; that is, it will show that the stone always falls in the same place on the ship, whether the ship is standing still or moving with any speed you please. Therefore, the same cause holding good on the Earth as on the ship, nothing can be inferred about the Earth's motion or rest from the stone falling always perpendicularly to the foot of the tower. (Drake, *Dialogue*, 126–28, 144–45)

Galileo has refuted the Aristotelian argument against motion of the Earth but has not proven motion of the Earth: neither motion nor rest can be inferred from the experiment.

Experiment has enjoyed a prominent role in the dialog up to this point, but that is about to change. Galileo was even more adept at using reason to demolish his opponents:

Simplicio: If you had referred me to any other agency than experiment, I think that our dispute would not soon come to an end; for this appears to me to be a thing so remote from human reason that there is no place in it for credulity or probability.

Salviati: For me there is, just the same.

Simplicio: So you have not made a hundred tests, or even one? And yet you so freely declare it to be certain? I shall retain my incredulity, and my own confidence that the experiment has been made by the most important authors who make use of it, and that it shows what they say it does.

Salviati: Without experiment, I am sure that the effect will happen as I tell you, because it must happen that way; and I might add that you yourself also know that it cannot happen otherwise, no matter how you may pretend not to know it—or give that impression. But I am so handy at picking people's brains that I shall make you confess this in spite of yourself . . . If only Simplicio is willing to reply to my interrogation, I cannot fail . . .

Now as to that stone which is on top of the mast; does it not move, carried by the ship, both of them going along the circumference of a circle about its center? And consequently is there not in it an ineradicable motion, all external impediments being removed? And is not this motion as fast as that of the ship?

Simplicio: All this is true, but what next?

Salviati: Go on and draw the final consequence by yourself, if by yourself you have known all the premises.

Simplicio: By the final conclusion you mean that the stone, moving with an indelibly impressed motion, is not going to leave the ship, but will follow it, and finally will fall at the same place where it fell when the ship remained motionless . . .

This argument is really very plausible in appearance, but actually it is offset by a difficulty which is hard to overcome. You have made an assumption throughout which will not lightly be granted by the Peripatetic school, being directly contrary to Aristotle. You take it as well known and evident that the projectile when separated from its origin retains the motion which was forcibly impressed upon it there . . .

So the rock cannot follow the motion of the boat either through any force impressed upon it . . . and therefore it will remain behind.

Salviati: How many propositions I have noted in Aristotle (meaning always in his science) that are not only wrong, but wrong in such a way that their

diametrical opposites are true, as happens in this instance! (Drake, *Dialogue,*
145, 148–50, 153–54)

Galileo proceeded to ridicule the Aristotelian alternatives to impressed motion.
Destroying Aristotelian physics and replacing it with a new physics hospitable
to motion of the Earth was a major part of the Copernican revolution, and necessary for its acceptance.

Galileo's telescopic discoveries, logical implications of the Copernican
system, and the principle of *plenitude,* which interpreted any unrealized
potential in nature as a restriction of the Creator's power, all combined to
encourage belief in a plurality of worlds. John Wilkins, a major figure in
the establishment of the Royal Society of London and its first secretary,
argued the case in his 1638 book *The Discovery of a New World: or, a
Discourse tending to prove, that it is probable there may be another habitable
World in the Moon.* Although there was no direct evidence of lunar inhabitants, Wilkins guessed that there were some inhabitants. Why else would
Providence have furnished the Moon with all the conveniences of habitation
shared by the Earth?

Persistent rumors that England intended to colonize the Moon further spread
the idea of a plurality of worlds and the new Copernican astronomy, although the
supposed plan was to send colonists to an empty Moon, not to subjugate indig-
enous lunar inhabitants. The rumors were perhaps begun by Wilkins's quotation
of the astronomer Johannes Kepler, who had predicted that as the art of flying
was developed, the successful nation would transplant a colony to the Moon.

The new Copernican astronomy also spread to a wider audience through
political and social criticism. The organization of lunar inhabitants became
either the model of a perfect society or the reflection of all the vices of the
Earth's society. This fictional aspect, not incidentally, furnished writers some
protection against outraged monarchs, much as the fiction that scientific
theories were merely hypothetical had freed them from theological oversight.
Samuel Colvil, in his 1681 *The Whigs Supplication,* described what one might
see through the telescope:

> If he once level at the Moon,
> Either at midnight or at noon,
> He discovers *Rivers, Hills,*
> *Steeples, Castles* and *Wind-Mills,*
> *Villages* and *fenced Towns,*
> With *Foussies, Bulwarks,* and *great Guns,*
> Cavaliers on horse-back prancing
> Maids about a may-pole dancing
> Men in Taverns wine carousing,
> Beggars by the high-way loafing,
> Soldiers forging ale-house brawlings,
> To be let go without their lawings;
> . . .

Young wives old husbands horning,
Judges drunk every morning,
Augmenting law-fruits and divisions,
By *Spanish* and *French* decisions;
Courtiers their aims missing,
Chaplains widow-ladies kissing;
Men to sell their lands itching,
To pay the expenses of their kitching,
Physicians cheating young and old,
Making both buy death with gold. (Hetherington, *Man Society and the Universe*)

Once the barrier in human imagination against extraterrestrial life was breached by imaginary lunar inhabitants, the concept of a plurality of worlds quickly spread to the planets, and then beyond our solar system to other planets circling other suns. In 1686 the French astronomer, mathematician, and writer Bernard Fontenelle published his *Entretiens sur la pluralité des mondes* (*Conversations on the Plurality of Worlds*). The book was an instant best-seller, made it onto the Catholic index of prohibited books, and continues to be read today. Fontenelle may have been inspired by Wilkins's book on a habitable world in the Moon, which was published, in French translation, in Fontenelle's hometown in 1655 (two years before Fontenelle's birth).

Fontenelle's book describes his imagined evening promenades in a garden with a lovely young marquise. Naturally, their conversation turned to astronomy. On the second evening, the hero explained to his eager and enthusiastic companion that the Moon was *une terre habite,* although an absence of atmosphere might change that conclusion. On the third night, he continued with the idea that the planets were also inhabited. On the fifth and final night, he argued that the fixed stars were other suns, each giving light to their own worlds. What a romantic flirtation!

Fontenelle was not explicitly a feminist. He did, however, take seriously women's intellectual ambitions. In this he was not typical of male intellectuals of his time. In 1672 Molière in his comedy *Les Femmes Savantes* had mocked women for trying to better themselves, for involving themselves in anything other than trivial, mindless pursuits—damned if they did think, and damned if they did not. Fontenelle helped open a market for women readers, perhaps even more in England than in France. In 1713 an English newspaper described a mother and her daughters making jam while reading Fontenelle aloud to each other. From passive consumers of scientific information to active participants in scientific investigations would be the next step for women. Seeking origins of the feminist social revolution in the Copernican scientific revolution is a tenuous but not entirely implausible quest.

Above all, the Copernican revolution was a revolution in our understanding of our place and meaning in the universe. The revolution saw a historical progression from belief in a small universe with humankind at its center to a larger, and eventually infinite, universe with the Earth not in the center. The physical geometry of our universe was transformed from geocentric and homocentric to heliocentric, and eventually to a-centric. The psychological change was no less. We no longer command unique status as residents of the center of the universe, enjoying our privileged place. Nor are we likely the only rational beings in the universe. One might even question whether a good God sent an Adam and an Eve and a Jesus Christ only to us, or to every planet.

Galileo, Science, and Religion

Galileo's support of Copernican astronomy culminated in a clash with church authorities so dramatic that it is the foundation of the most widely held stereotype regarding the relationship between science and religion: automatic antagonism and unavoidable war. The conflict between Galileo and the Catholic Church, however, was far from inevitable.

Initially, Aristotelian philosophers in Italian universities opposed Galileo. According to Galileo, his science was "in contradiction to the physical notions commonly held among academic philosophers," and they had "stirred up against me no small number of professors . . . " (Drake, *Discoveries*, 176).

They were eager to enlist the Church on their side. Galileo complained that they "hurled various charges and published numerous writings filled with vain arguments, and they made the grave mistake of sprinkling these with passages taken from places in the Bible which they had failed to understand properly . . . " (Drake, *Discoveries*, 176). They had "resolved to fabricate a shield for their fallacies out of the mantle of pretended religion and the authority of the Bible" (Drake, *Discoveries*, 177). A few individual priests were induced to charge that the motion of the Earth was contrary to the Bible.

At around the same time, a church official remarked that the Bible tells us how to go to heaven, not how the heavens go. Wisely, the Church was not eager to enter a scientific dispute.

Galileo attempted to win the Church to his side and silence objections to Copernican astronomy based on scripture. Saint Augustine earlier had suggested that no scientific doctrine should ever be made an article of faith, lest some better-informed heretic exploit misguided adherence to a scientific doctrine to impugn the credibility of proper articles of faith. Galileo cited Saint Augustine's warning. It was good advice for an era in which new telescopic observations were being made almost nightly.

Galileo also appealed to the authority of Saint Augustine in support of the thesis that no contradiction can exist between the Bible and science when the Bible is interpreted correctly. Galileo acknowledged "that it is very pious to say and prudent to affirm that the holy Bible can never speak untruth—whenever its true meaning is understood" (Drake, *Discoveries*, 181). But for "discussions of physical problems we ought to begin not from the authority of scriptural passages, but from sense-experiences and necessary demonstrations" (Drake, *Discoveries*, 182).

Galileo's position may sound sensible now, but he was out of step with his time. The Counter-Reformation then demanded tight control over Church doctrine, the better to counter Protestants.

In 1616 Pope Paul V submitted the questions of the motion of the Earth and the stability of the Sun to the official qualifiers of disputed propositions. It is not known why the pope acted.

(Continued)

(Continued)

Galileo expected the qualifiers to read the Bible metaphorically. Instead, they read it literally, and found both the motion of the Earth and the stability of the Sun false and absurd in philosophy. They did not rule on the truth of Copernican astronomy, on its agreement with nature. They did rule that the motion of the Earth was at least erroneous in the Catholic faith, and that the stability of the Sun was formally heretical.

The qualifiers had exceeded their authority because only the pope or a Church Council could decree a formal heresy. The pope ignored the finding.

The Congregation of the Index issued an edict forbidding reconciliation of Copernicanism with the Bible and assertion of literal truth for the forbidden propositions. One passage about scriptural interpretation and passages calling the Earth a star (implying that it moved like a planet) were ordered removed from Copernicus's *De revolutionibus*. Catholics could still discuss Copernican astronomy hypothetically, and little damage had been done to science.

At a meeting with Church officials, Galileo was instructed no longer to hold or defend the forbidden propositions: the motion of the Earth and the stability of the Sun. Had Galileo resisted, the Commissary General of the Inquisition was prepared to order him, in the presence of a notary and witnesses, not to hold, defend, or teach the propositions in any way, on pain of imprisonment. Galileo did not resist, but the Commissary General may have read his order anyway. It appears in the minutes of the meeting, unsigned and unwitnessed. Galileo may have been advised to ignore the unauthorized intervention. Subsequent rumors that Galileo had been compelled to abjure caused him to ask for—and he received—an affidavit stating that he was under no restriction other than the edict applying to all Catholics.

In 1623 a new pope was chosen. Urban VIII was an intellectual, admired his friend Galileo, granted him six audiences in 1624, and encouraged him to write a book on Copernican astronomy. The book, Urban hoped, would demonstrate that the Church did not interfere with the pursuit of science, only with unauthorized interpretations of the Bible.

Galileo's *Dialogue* on world systems was published in 1632, with Church approval. Urban, however, became angry when he found his own thoughts attributed to the Aristotelian representative who lost every argument in the *Dialogue*. (It wasn't prudent to anger Urban; one summer he had all the birds in the papal gardens in Rome slaughtered because the noise they made distracted him from his work.)

Also, the timing was unfortunate for Galileo. His book appeared in a climate of heightened suspicion, even paranoia. A Spanish cardinal had criticized Urban for interfering in a political struggle, and Urban had responded with a purge of pro-Spanish members of his administration, including the secretary, who, coincidentally, had secured permission for printing the *Dialogue*.

Galileo was called to Rome, charged with contravening the (unsigned and unauthorized) order of the Inquisition not to hold, defend, or teach the Copernican propositions in any way. Galileo produced his affidavit, signed and dated, stating that he was under no restriction other than the edict applying to all Catholics. Nonetheless, he was found guilty and compelled to abjure, curse, and detest his errors and heresies. Henceforth, even hypothetical discussion of Copernican astronomy was heresy for Catholics.

The Galileo fiasco has long been an embarrassment for the Catholic Church. In 1978 Pope John Paul II acknowledged that Galileo's theology was sounder than that of the judges who condemned him. In 1992 the pope set up a special committee to reexamine the Galileo case and offered an official apology for Galileo's sentence.

Galileo's entanglement with religious authority resulted in the stereotype of warfare between science and religion. This stereotype is a vast oversimplification, as most stereotypes are.

Faith in an anthropocentric universe lay shattered, leaving humankind's relationship with God uncertain. John Donne's 1611 poem *The Anatomy of the World,* with its opening line the "new Philosophy calls all in doubt," and later, "all Relation: Prince, Subject, Father, Son, are things forgot," refers to Christian morality as well as the physical locations of the Sun and the Earth.

The idea of a "Great Chain of Being," the manifestation in the world of God's thought, would soon be among things forgotten in the intellectual turmoil of the Copernican revolution. The chain linked God to man to lifeless matter in a world in which every being was related to every other in a continuously graded, hierarchical order. Every entity had its spot in the hierarchy. Governmental order reflected the order of the cosmos. Belief in the great chain of being precluded the possibility of evolution. Social mobility and political change were crimes against nature. All this began to change with Copernicus.

Astronomers still sought, however, as they had for thousands of years, to explain the apparently irregular motions of the planets, the Sun, and the Moon with combinations of uniform circular motions. Breaking the circle is the subject of the next chapter.

16

BREAKING THE CIRCLE

The Copernican revolution, a turning point in human thought, saw Aristotelian physics and the hierarchical universe swept away. Yet working astronomers continued to save the phenomena with geometrical models. Their revolution would occur only after the ancient assumption of uniform circular motion was destroyed. Observations by Tycho Brahe interpreted by Johannes Kepler would accomplish the deed.

Tyge Brahe, now customarily referred to by his first name in its Latin version, *Tycho*, was born in 1546. Before he was two years old, he was abducted by his uncle Jörgen Brahe. Tycho later wrote that his uncle "without the knowledge of my parents took me away with him while I was in my earliest youth" (Thoren, *Lord of Uraniborg*, 4). Tycho's father had more sons and eventually gave his permission for Tycho to be raised by the childless Jörgen. The expectation that Tycho would inherit Jörgen's wealth likely helped decide the matter.

The Brahe family was prominent in the governing of Denmark, and Tycho's brothers prepared themselves for government careers. For them, learning Latin, the scholarly language of Europe, would have been a waste of time. Tycho, however, encouraged by Jörgen's brother-in-law, Peder Oxe, began studies at the University of Copenhagen in 1559. Impressed by successful predictions of an eclipse, Tycho became interested in astronomy.

In 1562 Tycho traveled with a tutor to the University of Leipzig. The tutor was supposed to keep Tycho focused on studying law. Tycho, however, made observations while his tutor slept. He found that astronomical tables were inaccurate and resolved to improve them. To do so, he would need better instruments.

Shortly after Tycho's return to Denmark, in 1566, Jörgen and the king fell off a bridge near the royal castle at Copenhagen. Probably both had been drinking. Perhaps the king fell in first, and Jörgen jumped in to rescue him. Jörgen did not recover from his carousing and icy dunking but instead died of pneumonia. He had not yet completed making Tycho his legal heir, so his wife inherited instead.

Tycho went abroad again. In Rostock he lost the bridge of his nose in a duel. The two student protagonists of the duel had competed in studying mathematics, and the duel may have been over which of them was the more skilled mathematician. Thereafter, Tycho carried a little box of adhesive salve to hold on a nosepiece, supposedly an alloy of gold and silver. When Tycho's tomb was opened in 1901, the nasal opening of his skull was found to be rimmed with green, an indication of exposure to copper. Perhaps Tycho had a lighter-weight copper nosepiece for everyday wear?

Tycho returned to Denmark in 1567 but soon left on another trip abroad. Oxe, now lord high steward of Denmark, arranged for Tycho to be named to the next vacant cannonry at Roskilde Cathedral. Tycho anticipated living on the high income of a nobleman and at the same time pursuing a career as an astronomer and scholar—a career barred by custom to noblemen and not usually subsidized by the crown.

In Augsburg for much of 1569, Tycho inspired a wealthy humanist to pay for construction of a huge wooden quadrant on his country estate. It had a 19-foot radius, and its arc was marked off in sixtieths of a degree. Before Tycho could make many observations, however, he was called back to Denmark. His father was dying.

After the Brahe estate was settled, in 1574, Tycho was independently wealthy and could do whatever he wanted. Two of his brothers contracted marriages; Tycho began planning another trip. He already had a mate, but she was not a noblewoman, and they could not be formally married. An aristocratic marriage would have involved Tycho in courtly activities and taken time away from scholarly studies.

On the night of November 11, 1572, returning from his alchemical laboratory to eat supper, Tycho noticed an unfamiliar starlike object in the constellation Cassiopeia. If nearby, it would appear to shift its position with respect to the background stars. But its parallax, or angle of view, did not change from night to night. Evidently Aristotle had erred when he decreed there could be no change in the heavenly spheres beyond the Moon. Those who dismissed this implication of Tycho's discovery were in turn dismissed by him as thick wits and blind watchers of the sky. Publication of the discovery made Tycho famous, and he resolved to make astronomy his life's profession.

Tycho thought about moving to Basel, far from courtly distractions. King Frederick II, eager to retain credit and glory to himself and to Denmark for Tycho's work, offered Tycho the island of Hven and money for building and maintaining a proper establishment. Generous annual grants amounted to approximately 1 percent of the kingdom's total revenue. (In comparison, NASA's 1999 budget of $14 billion was 0.8% of total federal expenditures and 0.2% of gross domestic product.) Tycho also received rents from some 40 farms on the island and two workdays per week from each farm. He built a manor and an observatory named Uraniborg, also an underground observatory,

Figure 16.1: The Great Quadrant at Augsburg. Tycho recognized the need for systematic observations night after night with instruments of the highest accuracy obtainable. His wooden quadrant constructed in Augsburg in 1569 had a 19-foot radius and arc marked in sixtieths of a degree. It was very accurate, but also cumbersome, requiring many servants to align it, and permitting only a single observation nightly.

Source: From Tycho Brahe, *Astronomiæ instauratæ mechanica* [Instruments for the restoration of astronomy]. Noribergae [Nürnberg]: apud L. Hvlsivm [by Levinus Hulsius], 1602. 107 unnumbered pages, woodcut and engraved illustrations, map, plans.

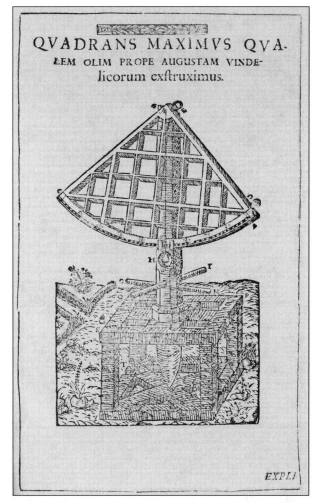

Figure 16.2: Tycho Brahe's Nova Stella or New Star of 1572. From Tycho Brahe, *De nova et nullius ævi memoria prius visa Stella, iam pridem Anno à nato CHRISTO 1572, mense Novembri primum Constecta, contemplatio mathematica.*

Tycho's new star was a supernova, a star that explodes and increases hundreds of millions of times in brightness. He marked it number I in the constellation Cassiopeia. Edgar Allan Poe had Tycho's new star in mind when he wrote his poem *Al Aaraaf* about a new star whose glory swelled upon the sky and whose glowing Beauty's bust beneath man's eye. Courtesy Copenhagen University Library.

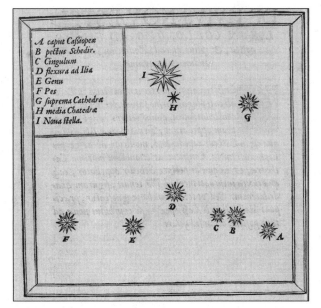

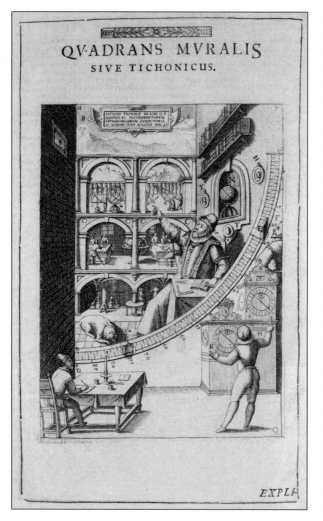

Figure 16.3: Tycho Brahe's Great Mural Quadrant at Uraniborg. The quadrant, completed in 1582, was forged of brass, 2 meters in radius, 13 millimeters broad, 5 millimeters thick, and marked in sixths of a minute.

Tycho points to the heavens. One assistant looks through the back sights; another reads off the time; a third records the data. The dog at Tycho's feet is a symbol of sagacity and fidelity. In the background is a representation of the main building with an alchemical laboratory in the basement, the library with its great globe and space for making calculations on the main floor, and above instruments for observing. In a niche in the wall behind Tycho is a small globe, automated to show the daily motion of the Sun and the Moon, and also phases of the Moon. On either side of the small globe are portraits of King Frederick II and Queen Sophia. Their financial generosity made the whole thing possible. At the top of the engraving is a landscape and setting sun.

Source: From Tycho Brahe, *Astronomiæ instauratæ mechanica* [Instruments for the restoration of astronomy]. Noribergae [Nürnberg]: apud L. Hvlsivm [by Levinus Hulsius], 1602. 107 unnumbered pages, woodcut and engraved illustrations, map, plans.

Stjerneborg, with more stability than his castle floors provided, large instruments for observing stellar and planetary positions, a paper mill and printing press, and other necessities for self-sufficiency.

The new star of 1572 was the first great astronomical event of Tycho's lifetime; the comet of 1577 was the second. Tycho showed that it was above the Moon, moving through regions of the solar system previously believed filled with crystalline spheres carrying around the planets. Aristotle's distinction between the corrupt and changing sublunary world and the perfect immutable heavens was shattered.

Although his observations contradicted Aristotelian cosmology and the medieval world view, Tycho's own conception of the universe differed little. Both Aristotelian physics and common sense dictated that the Earth did not move through the heavens, and Tycho kept the Earth in the center of the universe. In his world system, the Sun and the Moon circled the Earth and the outer sphere of the stars rotated about the Earth at its center every 24 hours. Where Tycho's

Figure 16.4: The Tychonic World System. In Tycho's system, the Sun and the Moon circle the Earth (the black dot at the center). The outer sphere of the stars rotates about the Earth at its center every 24 hours. Mercury, Venus, Mars, Jupiter, and Saturn are in circular orbits centered on the Sun.
Source: Engraving from Tycho Brahe, *De mundi aetherei recentioribus phaenomenis,* 1588. Courtesy Copenhagen University Library.

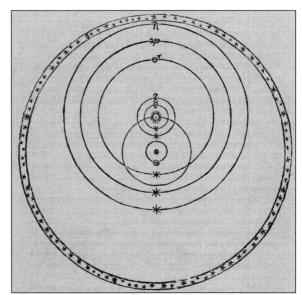

Mental Exercise: Tycho's System

List astronomical phenomena that the Tychonic world system accounts for automatically but that the Ptolemaic system does not. Explain how the Tychonic system accounts for the phenomena.

system differed from Ptolemy's was in placing Mercury, Venus, Mars, Jupiter, and Saturn in circular orbits centered on the Sun rather than on the Earth.

In departing from the ancients and placing the planets in orbits centered on the Sun, Tycho gained many of the geometric advantages of the Copernican system without displacing the Earth from the center of the universe, nor placing it in motion. The motions of the Sun and planets relative to the Earth are mathematically and observationally equivalent in the Tychonic and Copernican models. Both systems would accommodate the phases of Venus, after they were observed by Galileo. Tycho's system became a popular alternative for astronomers forced to give up the Ptolemaic system but not ready to accept the Copernican system. They don't seem to have been bothered by, or even to have been aware of, the narrowed distinction between the Sun and the planets and the diminished status of the Sun in Tycho's system.

Tycho became something of a scandal at court for forcing his islanders to work without pay and imprisoning them if they complained. The new king, Christian, crowned in 1596, admonished Tycho. Arrogant and domineering, Tycho did not receive admonishment graciously. Christian next slashed Tycho's

Figure 16.5: The Tychonic, Copernican, and Ptolemaic Systems Compared. Giambattista Riccioli was a Jesuit and thus prohibited from teaching the Copernican system. In the frontispiece of his 1651 *Almagestum novum* (*New Almagest*), Astrea, the Goddess of justice, is the winged angel holding the balance beam of truth. The intermediate Tychonic system outweighs the Copernican heliocentric system. The Ptolemaic geocentric system, found wanting in the balance, is discarded in the lower right corner. Courtesy Copenhagen University Library.

funding. After a brief stop in Copenhagen, presumably to give Christian a last chance to retain him for the glory of Denmark, Tycho sailed away with his instruments, printing press, and two dozen household servants in search of a new patron. After two years of wandering, Tycho found favor with Emperor Rudolph II of the Holy Roman Empire. Tycho set up his instruments in a castle near Prague. He died there two years later.

Legend has it that Tycho died from uremia due to a ruptured bladder after too large a meal. According to a note written in Tycho's observation logbook by an assistant, on October 13, 1601, Tycho drank a bit overgenerously at a banquet, felt some pressure on his bladder, but remained seated rather than commit a breach of etiquette. By the time he reached home he could no longer urinate. There followed excruciating pain, insomnia, and delirium before Tycho died on October 24. He may have had an enlarged prostate, although he wasn't all that old. A kidney stone is another possibility, but none were found when Tycho's body was exhumed in 1901. A more recent examination, in 1996, of strands of Tycho's beard found increased levels of mercury. This might be explained by exposure to mercury in his alchemical laboratory while trying to make gold. One of the hairs, however, seemingly had a high local concentration of mercury in the root, suggesting that Tycho ingested the mercury during the last day of his life. If so, did he take it as medicine or by mistake, or was he poisoned?

Johannes Kepler succeeded Tycho as Imperial Mathematician at the Prague court. Kepler was born in southern Germany in 1571, into what has been described as a family of impoverished, degenerate psychopaths. His father deserted the family early on, and his mother later was tried as a witch. Kepler would not have received an education had not the Dukes of Württemberg created outstanding Protestant universities to produce government administrators and clergymen to lead the reformation then raging in Germany. There was also a network of elementary and secondary schools, and scholarships for children of the faithful poor.

At the University of Tübingen, Kepler received a classical education, including mathematics and astronomy. His professor, Michael Maestlin, was a Copernican, perhaps the only Copernican professor at that time. Kepler seemingly was destined for the clergy, to become a Lutheran minister, and he proceeded to the theological school.

In 1593 the mathematician in the

Figure 16.6: Portrait of Tycho Brahe. Frontispiece from Tycho Brahe, *Astronomiæ instauratæ mechanica* [Instruments for the restoration of astronomy]. Noribergae [Nürnberg]: apud L. Hvlsivm [by Levinus Hulsius], 1602. 107 unnumbered pages, woodcut and engraved illustrations, map, plans.

Protestant school at Graz, the capital of the Austrian province of Styria, died, and the school sought a recommendation from Tübingen for a replacement. Perhaps the Tübingen professors wanted to be rid of the querulous young Kepler, who had defended Copernicus in a public disputation. Maybe they hoped he would make a better teacher than a priest.

At Gratz on July 9, 1595, so Kepler later wrote, while drawing a figure on a blackboard he was suddenly struck so strongly with an idea that he felt he was holding the key to the secret of the universe. He had been showing his class how consecutive conjunctions of Jupiter and Saturn take place every 20 years and fall successively about one-third of the way around the sky. Then he realized that for an equilateral triangle, whose points fall exactly in thirds of the way around a circle,

> the radius of a circle inscribed in such a triangle is one half the radius of the circumscribed circle. The ratio between the two circles was, to the eye, exactly the same as that which is found between [the orbits of] Saturn and Jupiter, and the triangle is the first of the geometrical figures, just as Saturn and Jupiter are the first planets. I immediately tried [to determine] the second distance, between Mars and Jupiter, by means of a square, the third by means of a pentagon, the fourth by means of a hexagon. (Koyré, *Astronomical Revolution*, 141)

Kepler thought he had discovered a grand design, the cosmic order linking planetary radii with a progression of basic polygons.

But then it struck Kepler that there is no natural limit on the number of polygons. He wrote: "I realized, in fact, that if I wanted to continue the arrangement of the figures in this manner, I should never come to the Sun, and that I should never find the reason why there must be six planets [there were then six known planets] rather than twenty or one hundred" (Koyré, *Astronomical Revolution*, 142).

The necessity to explain why there were six planets, not any disagreement between Kepler's initial idea and observational data, drove him on almost immediately from basic polygons to regular solids:

> I then said to myself, I should obtain what I am seeking, if five figures, among the infinitely many, can be found for the size and relationships of the six spheres assumed by Copernicus. Then I went even further. What have plane figures to do with corporeal spheres? Clearly one should resort to solid bodies. Here, then, dear reader, is my discovery and the content of my short treatise. If this be told

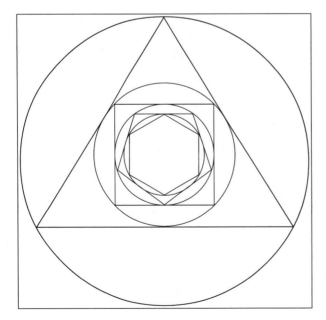

Figure 16.7: Basic Polygons between Circular Orbits. Kepler imagined a mathematical harmony involving planetary radii and polygons. He drew the first basic polygon, an equilateral triangle, with three sides, inside the circle of Saturn's orbit. Next, within this triangle, he drew Jupiter's circular orbit. Remarkably, the ratio between the two circles' radii was the same as that observed between the orbits of Saturn and Jupiter. Then Kepler placed the second basic polygon, a square, with four sides, inside Jupiter's orbit; and he placed Mars's orbit within the square. Then, he placed a pentagon, with five sides, between the orbits of Mars and Earth. Then, he placed a hexagon, with six sides, between the orbits of Earth and Venus, and so forth.

to anyone with the least knowledge of geometry, he will immediately think of the five regular solids and their relationship to their circumscribed and inscribed spheres. (Kayré, *Astronomical Revolution*, 142)

The coincidence could not be purely fortuitous, could it? There were six planets, five intervals between them, and five regular solids! Kepler proclaimed the discovery in his *Mysterium cosmographicum (Cosmic Mystery)* in 1596.

Kepler's emphasis on mathematical harmony might have been expected to appeal to Neoplatonists and Neopythagoreans. And his delight in explaining the existence of six planets as the natural and inevitable consequence of theory rather than merely an arbitrary parameter revealed by observation might have been expected to resonate with astronomers sharing Copernicus's philosophical values. Yet Kepler's model of nested polyhedra was not well received.

Kepler sent copies of his *Mysterium cosmographicum* to leading astronomers. One responded that he preferred to stick to "plain astronomical speculations" and not "endeavor to pry into more secret principles of nature with physical or metaphysical reasons," which he believed "cannot be of use to the astronomer in almost any way" (Voelkol, *Kepler's* Astronomia Nova, 79). Not speculative a

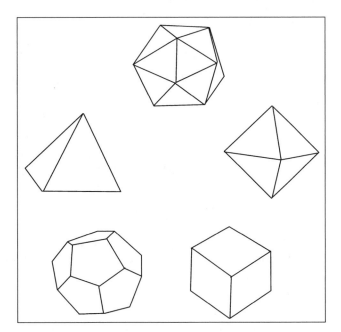

Figure 16.8: The Five Regular Solids. There are five regular (or perfect or Platonic or Pythagorean) solids; no more, no less. All sides are equal, all angles are the same, and all faces are identical. They are the tetrahedron (4 triangular sides), cube (6 square sides), octahedron (8 triangular sides), dodecahedron (12 pentagonal sides), and icosahedron (20 triangular sides).

About the Earth's circle Kepler circumscribed a dodecahedron; then around the dodecahedron he drew Mars' circle; then about Mars' circle he circumscribed a tetrahedron; around the tetrahedron he drew Jupiter's circle; about Jupiter's circle he circumscribed a cube; and he enclosed the cube with Saturn's circle. Within the Earth's circle, Kepler inscribed an icosahedron; in it he inscribed Venus' circle; in this circle, he inscribed an octahedron; and he inscribed Mercury's circle inside the octahedron.

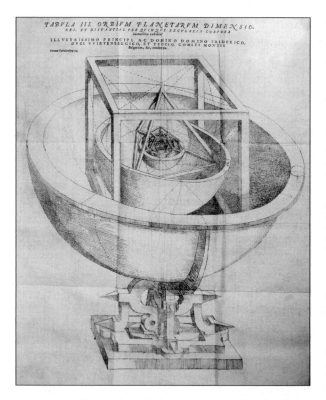

Figure 16.9: Kepler's Model of the Solar System; Close-up of the Model.

Image copyright History of Science Collections, University of Oklahoma Libraries.

priori reasoning from theory or principles without prior observations, but a posteriori reasoning from observation to theory was the proper way to determine the sizes of the planets' orbits. Kepler's critic continued: "What does it matter whether or not they coincide with the regular bodies? If he [Kepler] truly thinks that after this kind of principle has been proposed . . . calculations are to be bound to it, I think this is done in preposterous order" (Voelkol, *The Composition of Kepler's*, 79).

Kepler's theory was rightly perceived as a mystical a priori speculation. It had originated more in his imagination than in observation. His order of work was preposterous. The proper procedure decreed for astronomers (however much they believed a priori in uniform circular motion) was to derive the distances of the planets a posteriori from observations, not a priori from the geometry of regular solids.

Probably as damning, Kepler had ventured outside the astronomical tradition of saving the phenomena with uniform circular motions, while leaving questions about causes to physicists. Kepler was more a mathematical mystic and a physicist than a geometrical astronomer. He pried into secret principles of nature with physical or metaphysical reasons. Years later, in 1616, Maestlin would write to Kepler: "Concerning the motion of the Moon you write you have traced all the inequalities [deviations from uniform circular motion] to physical causes; I do not quite understand this. I think rather that here one should leave physical causes out of account, and should explain astronomical matters only according to astronomical method with the aid of astronomical, not physical, causes and hypotheses. That is, the calculation demands astronomical bases in the field of geometry and arithmetic" (Voelkol, *Kepler's* Astronomia Nova, 69).

Tycho Brahe, another recipient of Kepler's *Mysterium cosmographicum*, also thought that Kepler should look for cosmological harmonies a posteriori, after the motions were ascertained with precision, not a priori. Tycho added: "If more accurate measurements . . . of the kind collected over many years I have at hand, were applied, they could furnish a more accurate test" (Voelkol, *Kepler's* Astronomia Nova, 84). Tycho had the data Kepler needed were he ever to command the attention of astronomers with a convincing and compelling proof.

The most important result following from Tycho's receipt of Kepler's *Mysterium cosmographicum* was a vague invitation to visit Tycho someday (and somewhere, because he was then wandering about Europe in search of a new patron). However nebulous, Tycho's invitation would be particularly welcome after Archduke Ferdinand of Hapsburg decided to cleanse his Austrian provinces of Lutheran heresy. Kepler's school was closed in 1598, and all Lutheran schoolmasters and clergy were ordered to leave in eight days or forfeit their lives. The archduke was pleased with Kepler's discoveries, however, and soon allowed him to return. Still, talk of torturing and burning heretics must have worried Kepler.

Tycho was now settled near

Figure 16.10: *Stars.* Engraving by M. C. Escher, 1948. © 2005 The M.C. Escher Company-Holland. All rights reserved. www.mcescher.com.

Prague, a result which Kepler, who could not have afforded a trip to distant Denmark, would later attribute to Divine Providence. By happy chance, one of Emperor Rudolph's councilors was in Graz about to return to Prague, and he agreed to take Kepler with him. Otherwise, Kepler might not have been able to afford even this relatively short trip. They departed for Prague on January 1, 1600.

Kepler wanted to test his own theory. Tycho, however, had hired Kepler to test Tycho's theories. Not trusting Kepler, Tycho only slowly trickled out observations to Kepler, and only for Mars. That Tycho set him to work on Mars's orbit Kepler also attributed to Divine Providence. He wrote: "For Mars alone enables us to penetrate the secrets of astronomy which otherwise would remain forever hidden from us" (Koestler, *Sleepwalkers*, 315). The motions of the other planets are so close to circular that even Tycho's data would not have revealed their noncircularity. Mars's orbit alone deviates sufficiently from a circle that Tycho's observations would force Kepler from an eccentric circle to an oval and eventually to an ellipse.

Observations of unprecedented accuracy were crucial, but they were not the only element in Kepler's work on Mars. He sought a physical mechanism to explain the planet's motion. He aimed to show that the celestial machine was not a divine organism but rather a clockwork with all its movements carried out by means of a single, simple magnetic force. Moreover he would demonstrate this physical conception through calculation and geometry. Kepler seems to have thought that the force emanating from the Sun driving the planets was analogous to magnetism, or even magnetism itself. William Gilbert, royal physician to both Elizabeth I and James I in England, had published *De magnete,* his study of the magnet and magnetic bodies and that great magnet the Earth, in 1600.

The first result of Kepler's obsession with physical cause was that he made his calculations with the center of Mars's orbit not at the center of the Earth's orbit but at the Sun. A mathematical point, Kepler reasoned could not move and attract a heavy object. The Sun, he believed, was the physical center and cause of the planets' motions. Centering Mars's orbit on the Sun also better fit observations, but Kepler's initial motivation was more philosophical. Justification of a scientific theory relies on reason, while discovery often is a product of human emotion. Kepler's obsession with physical cause also discouraged him from employing implausible nonphysical devices such as epicycles and equants (but eccentric circles were acceptable) in his many attempts to fit an orbit for Mars to Tycho's data.

Tycho's sudden death in 1601 liberated Kepler to pursue his own interests. Tycho's death also left Kepler with the data he coveted. Previous astronomers had observed the planets only at critical moments, such as opposition. In contrast, Tycho's more omnivorous observing program caught the planets at more points in their orbits. When Kepler posed new questions, Tycho's data would sometimes yield an answer. To a correspondent, Kepler confessed that he had quickly taken advantage of the absence of Tycho's heirs and taken Tycho's observations under his care, perhaps even usurping them.

Kepler still had to deal with conventional astronomers' criticisms. There is more to the advance of science than new observations and new theories. Ultimately, people must be persuaded. Kepler's *Astronomia nova (New Astronomy)* of 1609 is not the stream-of-consciousness narrative, the aimless wandering full of failures and excursions up blind alleys in his "war on Mars" that it appears to be. Rather, the book is a cleverly composed argument. Kepler had learned well his lesson from the negative reception given his *Mysterium cosmographicum.* If astronomers wanted a posteriori, he'd give them a posteriori—or at least what looked like it. The title page of his *Astronomia nova* read "A NEW ASTRONOMY Based on Causation or A PHYSICS OF THE SKY, derived from Investigations of THE MOTIONS OF THE STAR MARS, Founded on Observations of the Noble TYCHO BRAHE."

First, Kepler feigned a good faith effort to find a theory in the classical form of uniform circular motions. He investigated Mars's orbit "in imitation of the ancients." He failed, and in failing, he convinced his readers that the problem was intractable. Only then did he ostensibly look to physical cause as the only approach that could succeed: "After trying many different approaches to the reform of astronomical calculations, some well trodden by the ancients and others constructed in emulation of them and by their example, none other could succeed than the one founded upon the motions' physical causes themselves" (Voelkol, *Kepler's* Astronomia Nova, 225).

Kepler tried an oval orbit. Discrepancies between Tycho's observations and circular and oval orbits were equal and opposite. An elliptical orbit fit in between the circular and the oval. Thus what would become known as Kepler's first law: the planets move in elliptical orbits with the Sun at one focus. Kepler had expected to solve the problem of the orbit of Mars in a few weeks; instead, the solution consumed six years of intense labor.

How Kepler could so casually jettison circular orbits and two thousand years of tradition for the ellipse is a question that worries historians, if not astronomers who care only that the decision was proved right by subsequent events. Kepler was not an astronomer inextricably enmeshed in regular circular motions, but more of a mathematician. He was well acquainted with Apollonius's book on conic sections, which had been recovered along with other Greek classics in

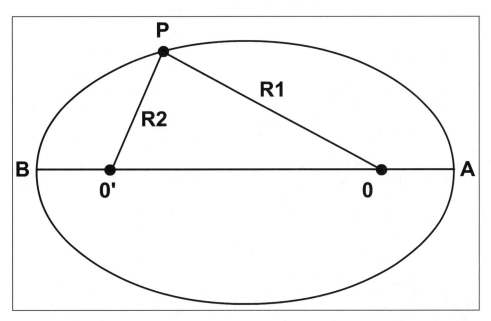

Figure 16.11: Ellipse. An ellipse is defined as the collection of all points for which the sum of R1 and R2 is a constant. Also, an ellipse is the closed curve generated by the point *P* (the locus of the points *P*) moving in such a way that the sum of its distances from two fixed points, the foci *(O' and O)*, is a constant. The planet is at *P*. The Sun is at one focus point, either *O'* or *O*. *AB* is the major axis. The minor axis is perpendicular to *AB* and intersects it at its midpoint, midway between *O'* and *O*.

The amount that an ellipse differs from a circle is called its *eccentricity*. This value is between zero and one. An ellipse with zero eccentricity (major axis *a* = minor axis *b*) is a circle. An ellipse with an eccentricity of one (minor axis *b* = 0) is an infinitesimally thin straight line, the same line as the major axis. The eccentricity of an ellipse is defined with mathematical precision in the following formulation, in which *a* is half the length of the major axis and b is half the length of the minor axis:

$$\frac{\sqrt{a^2 - b^2}}{a}$$

the early stages of the Renaissance. Indeed, Kepler explicitly contrasted the difficulty of his *Astronomia nova* to the *Conics* of Apollonius. The circle and the ellipse are both conic sections, as Kepler noted: "[T]he circle and the ellipse are from the same genus of figures" (Voelkol, *Kepler's* Astronomia Nova, 190). What, then, is so difficult or revolutionary in substituting one for the other in planetary theory? Neither geometrical figure was, for Kepler, any more natural or more perfect than the other.

Kepler's obsession with physical aspects may also have contributed to the surprising ease with which he abandoned uniform circular motion. Epicycles, for him, existed only in thought, not in reality. Nothing coursed through the heavens except the planets themselves.

In addition to possible philosophical reasons, there was also an immediate practical reason for Kepler to embrace the ellipse: its geometrical properties simplified his calculations. He had discovered another numerical harmony, that the line joining a planet to the Sun sweeps out equal areas in equal intervals of time, and it was easier to calculate the area for an elliptical orbit than for an oval orbit.

That the planets move faster the nearer they are to the Sun had already been cited by Copernicus as a celestial harmony. Kepler now found further harmony in a quantitative formulation of the relationship: equal areas in equal times. It was not immediately apparent, though, that this discovery would be hailed enthusiastically in textbooks four centuries later as Kepler's second law.

Very likely, thoughts about a force emanating from the Sun and attenuated by distance had helped direct Kepler to his law of equal areas. He believed

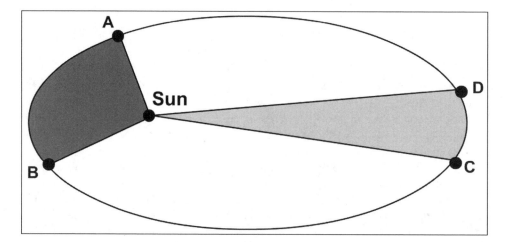

Figure 16.12: Kepler's Law of Equal Areas. The radius vector, the line from the Sun to the planet, sweeps out equal areas in equal times. If the areas of the two segments are equal, then the times for the planet to travel between points *A* and *B* and between points *C* and *D* are also equal. Thus the distance of a planet from the Sun is inversely related to its orbital velocity: as the distance increases, the velocity decreases.

that the Sun diffused motive power across the region containing the planets in the same way that God acted through the Holy Ghost. And consequently, it followed that motion was proportional to distance.

Kepler ended his *Astronomia nova* with an appeal. Arithmetic and geometry had triumphed in the battle against Mars. Now, Kepler wished to press the battle against the other planets: "Yet, I beseech Your Majesty to remember that money is the sinew of war, and to have the bounty to order Your treasurer to deliver up to Your general the sums necessary for the raising of fresh troops" (Kayré, *Astronomical Revolution*, 278). The emperor's treasurer had difficulty even finding funds to proclaim the victory over Mars. Finished in 1606, the *Astronomia nova* wasn't published until 1609.

Kepler's situation only worsened. The year 1611 was an altogether dismal and calamitous year. His salary was in arrears, as usual. War spread to Prague, bringing epidemics. Kepler's wife and one of his children died from diseases. And the emperor abdicated. The following year, Kepler was given the position of provincial mathematicus in Linz, the capital of Upper Austria. At least the Austrians were able to pay his salary. War eventually reached Linz, and in 1628 Kepler moved again, becoming court astrologer to the Duke of Wallenstein. Two years later, off attempting to collect money still owed him by the imperial treasury, Kepler became ill with a fever and died.

The years in Linz had been productive. Kepler published the *Harmonice mundi* (*Harmonies of the World*) on cosmic harmony in 1619; the *Epitome astronomiae Copernicanae* (*Epitome of Copernican Astronomy*) in three parts, between 1617 and 1621; and in 1627 the *Tabulae Rudolphinae* (*Rudolphine Tables*) of planetary positions, begun by Tycho under the patronage of Rudolph II nearly three decades earlier.

In both *Harmonies of the World* and *Epitome of Copernican Astronomy* the archetype of the universe remained the five regular solids. Nor were other harmonic consonances expressed by the Creator neglected. Kepler believed that God established nothing without geometrical beauty. Kepler compared the intervals between planets with harmonic ratios in music. He believed that the planetary orbits were governed by simple mathematical relations analogous to the mathematical relations discovered between harmonious musical tones by the Pythagoreans.

Among many propositions in the *Harmonies of the World* detailing various planetary ratios was the statement that the ratio of the mean movements of two planets is the inverse ratio of the $3/2$ powers of the spheres. Readers could scarcely have guessed that this particular harmony would later be singled out for acclaim as Kepler's third law while all the other numerical relationships in the book would be discarded as nonsense, tossed into the garbage can of history.

Kepler's third, or harmonic, law is now more usually stated as: the square of the period of time it takes a planet to complete an orbit of the Sun is proportional to the cube of its mean distance from the Sun. The law is also expressed as a ratio between two planets (*A* and *B*) going around the Sun (and also between

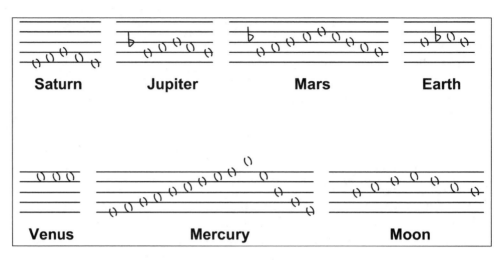

Figure 16.13: Kepler's Music of the Spheres. Kepler believed that the most wise Creator created harmonies between the planetary velocities, and that the harmonies are communicated to us by sound. Musical intervals correspond to ratios of the greatest and least velocities of each planet. Mercury and Mars with the most eccentric orbits, and thus greatest range of velocities, have the largest range of musical notes.

Plato had believed that through imitating the music of the spheres, humankind was returning to paradise, and the Roman Cicero wrote similarily that music enabled humankind to return to the divine region. Dante heard the music of the spheres as he ascended from purgatory to heaven, and Bach imagined that the sounds from his piano were a message from heaven. More recently, Einstein wrote that Mozart's music is so pure that it seems to have been present in the universe waiting to be discovered. Laws of physics, too, are preestablished harmonies with stunning symmetries waiting to be plucked out of the cosmos by someone with a sympathetic ear.

two satellites going around a planet): the ratio of the periods squared is equal to the ratio of the distances cubed

$$(\text{period A/period B})^2 = (\text{distance A/distance B})^3$$

How could anyone have ever noticed the correlation between distances squared and periods cubed? Two simplifications render the relationship more visible. First, if numbers are expressed in astronomical units and years, the distance and the period for the Earth are both 1. Second, Kepler's third law is a simple linear relationship between the logarithms of the planets' orbital periods and their distances from the Sun. Once Kepler began using logarithms, he would have seen his data in a new way, and his marvelous mathematical mind then could more easily have noticed the logarithmic harmony. That Kepler was thinking in terms of logarithms is suggested by his language: he wrote of the $^3/_2$ power, as one would if thinking logarithmically, rather than expressing his law in squares and cubes, as is now usually done.

We can guess the timing of Kepler's discovery from a comment in his *Harmonices mundi:* "Having perceived the first glimmer of dawn eighteen months ago, the light of day three months ago, but only a few days ago the plain

Figure 16.14: Log—Log Plot of Orbital Period versus Distance from the Sun.
John Napier, a Scottish nobleman, invented one of the most useful arithmetical concepts in all of science and published it in 1614 in his book *Mirifici logarithmorum canonis descriptio* (*Description of the Admirable Table of Logarithms*). Logarithms reduce the otherwise lengthy process of multiplication to simple addition. This might not seem like a big deal to owners of electronic calculators, but logarithms roughly doubled Kepler's productivity and his working lifetime.

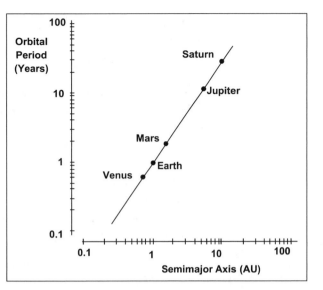

The logarithm L to the base 10 of the number X is defined by the equation $X = 10^L$. Thus for numbers between 0 and 10, the logarithm will be between 0 and 1; for numbers between 10 and 100, the logarithm will be between 1 and 2; and so forth. The logarithm of 1.3 is 0.113943, and the logarithm of 6.9 is 0.838849. To multiply 1.3 and 6.9, add their logarithms, obtaining 0.952792, which is the logarithm of the product of 1.3 and 6.9. Finding a number when its logarithm is known is called finding the *antilogarithm*. Napier compiled tables in which logarithms and antilogarithms could be looked up.

To square a number, which is the same as multiplying it by itself, either add the two identical logarithms together or multiply the logarithm by 2. To cube a number, multiply the logarithm by 3, and so forth. Division is performed by subtraction, comparable to multiplication by addition. Square and cube roots are calculated by division, comparable to squaring and cubing numbers by multiplication.

sun of a most wonderful vision—nothing shall now hold me back" (Koestler, *Sleepwalkers*, 393). The first dawn for Kepler's third law may have been when he first learned of logarithms, late in 1616. Broad daylight would have been March of 1618, when Kepler tested the third law but made errors in his calculations and set the idea aside. Full Sun was when Kepler took up again the calculations in May of 1618 and this time got them right.

Kepler was the first astronomer to seek universal physical laws based on terrestrial mechanics comprehending the entire universe quantitatively. He struggled to develop a philosophy or physics of celestial phenomena in place of the theology or metaphysics of Aristotle and in place of the traditional astronomers' fictitious geometry of epicycles and equants. Kepler asked questions others didn't; he sought answers where others didn't even see problems. Copernicus had noted that the motion of the heavenly bodies was circular because rotation was natural to a sphere; no further explanation was required. Nor, indeed, was explanation even a concern of Greek geometrical astronomers. They described *how* the heavens went; they didn't explain *why*. The physical nature and cause of orbital motion was, however, central to Kepler's concerns. His astronomy was revolutionary, or would have been had anyone paid any attention to it.

Brilliant as Kepler's work was, it won few adherents. The physical causes, the five regular solids, and the musical harmonies were all alien to traditional astronomy. Kepler's planetary theories commanded assent because of their unprecedented accuracy, but few astronomers adopted them. There was no Keplerian cosmology or world system as there were Ptolemaic, Copernican, and Tychonic world systems. In many ways, Kepler's work was a historical dead end.

Nor were Kepler's laws foundational in Isaac Newton's derivation of the inverse square law of gravity, as we will see in the next chapter. Kepler's observations did not lead to Newton's theory. Rather, establishment of the concept of universal gravitation would enshrine Kepler's three laws among the great achievements of science. Kepler's laws were important, however, in the acceptance of Newton's theory.

17

ISAAC NEWTON AND GRAVITY

In ancient Greek and medieval European cosmology, solid crystalline spheres provided a physical structure for the universe and carried the planets in their motions around the Earth. Then Copernicus moved the Earth out of the center of the universe, Tycho Brahe shattered the crystalline spheres with his observations of the comet of 1577, and Kepler replaced circular orbits with ellipses. How, now, could the planets continue to retrace their same paths around the Sun for thousands of years? The answer, encompassing both physical cause and mathematical description, came forth from Isaac Newton.

Newton's remarkable intellectual accomplishments include creation of the calculus, invention of the reflecting telescope, development of the corpuscular theory of light (its colored rays separated by a prism), and development of the principles of gravity and terrestrial and celestial motion. Newtonian thinking came to permeate not only the physical world but also intellectual fields, including politics and economics, where others were encouraged to seek universal natural laws of the sort Newton had found in physics and astronomy. Nor was eighteenth-century literature oblivious to the Newtonian revolution in thought. He changed history, perhaps more so than any other single person.

Newton's father, although a wealthy farmer and lord of his own manor, could not sign his name. Probably he would not have educated his son; his brother did not. Newton was born in 1642, the year Galileo died, three months after his father's death, premature, and so small that he was not expected to survive. When he was three, his mother remarried, and Newton was left in his grandmother's care. Difficult early years probably contributed to his difficult mature personality. He rejoined his mother seven years later, after his stepfather died, leaving his mother wealthy. Her family was educated, and she soon sent Newton off to school. He returned home at age 17 to learn to manage the family manor. But he was a disaster at rural pursuits, his mind lost in other thoughts. His mother's brother, a clergyman, urged her to send young Newton back to school to prepare for university. Newton entered Trinity College, Cambridge, in 1661. He would be appointed a fellow in 1667 and named Lucasian professor in 1669, at age 26. In later years, he would serve as warden of the mint and president of the Royal Society. He died in 1727.

At Cambridge, Newton initially studied Aristotelian physics. But around 1664, his notebooks reveal, he learned of the French scientist René Descartes. In his 1637 *Le discours de la méthode pour bien conduire sa raison et chercher la vérité dans les sciences* (*Discourse on the Method of Rightly Conducting Reason and Seeking Truth in the Sciences*), Descartes had begun with the famous phrase "Cogito ergo sum" (I think, therefore I am). From this certainty, he expanded knowledge, one step at a time, to include the existence of God, the reality of the physical world, and its mechanistic nature. Descartes' universe consisted of huge whirlpools, or vortices, of cosmic matter. Our solar system was one of many whirlpools, its planets all moving in the same direction in the same plane around a luminous central body. Planets' moons were swept along by the planets' vortices. All change in motion was the result of percussion of bodies; one object could act on another only by contact. Gravity was the result of celestial matter circulating about the Earth and pushing all terrestrial matter toward the Earth.

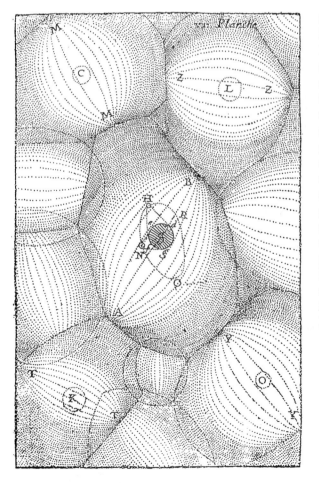

Figure 17.1: Whirlpools of Matter. Descartes, *Principia philosophiae*, 1644. Descartes speculated that God had divided matter into particles of the same size and imparted to them rotation around their centers and a propensity to rotate as a group around centers equidistant from each other. Distortion caused by unequal pressures of adjoining vortices reshaped circular orbits. Gravity's cause was pressure from celestial matter circulating in all directions about the Earth and tending to push objects toward the Earth. In the diagram, the Sun *(S)* is in the middle of the vortex *AYBM* and surrounded by a whirlpool of matter in which its planets circulate. Neighboring vortices each have a star at their centers. From René Descartes, *Principia philosophiae* (Leiden: Elzevier, 1644).

In contrast to Descartes' explanation of the phenomena of nature in terms of particles bouncing off each other, Newton would explain planetary motions as the result of a mysterious attraction at a distance. The path to his great achievement is almost as mysterious.

In 1665 the plague arrived in Cambridge, the colleges closed, and teachers and students fled to the countryside. There, a flash of insight supposedly gave Newton the idea of universal gravitation. So the legend goes, while musing in a garden, it occurred to him that the power of gravity causing an apple to fall to the Earth might also retain the Moon in its orbit. He did a quick calculation to check his idea, but because he had the wrong number for the radius of the Earth, his answer did not agree with his theory, and he put the theory aside.

Skeptical historians question the story of Newton and the apple, even though it was told by the great man himself. Newton's papers reveal that around 1666 he was thinking more in terms of the Moon fleeing from the Earth than being bound by the force of gravity to the Earth.

In 1679 the English scientist Robert Hooke wrote to Newton asking for his opinion about Hooke's hypothesis that the planetary motions might be compounded of a tangential motion and an attractive motion towards the central body. Years later, Hooke would accuse Newton of stealing and using this idea. Newton then remembered that he had also discussed the problem of determining the Heavenly motions upon philosophical principles with the architect Christopher Wren, in 1677. Wren confirmed this and stated that for many years he had had thoughts upon making out the planets motions by a composition of a descent towards the

Newton and the Apple

In the final year of his life, Newton told at least four different people how a sudden insight in 1666 supposedly gave him the idea of universal gravitation. An apple did not actually fall on his head, nor did he observe one to fall; he merely mused about apples falling. One recorded version of Newton's story reads:

> [W]hilst he [Newton] was musing in a garden it came into his thought that the power of gravity (which brought an apple from the tree to the ground) was not limited to a certain distance from the Earth but that this power must extend much farther than was usually thought. Why not as high as the Moon said he [Newton] to himself & if so that must influence her [the Moon's] motion & perhaps retain her in her orbit, whereupon he fell a calculating what would be the effect of that supposition . . . but being absent from books & taking the common estimate in use . . . his computation did not agree with his theory & inclined him then to entertain a notion that together with the force of gravity there might be a mixture of that force which the Moon would have if it was carried along in a vortex. . . . (Westfall, *Never at Rest*, 154)

Newton had used the wrong number for the radius of the Earth in his calculation, and thus the answer did not agree with his theory that the power of gravity which brought an apple from the tree to the ground was the same power that retained the Moon in its orbit. So he set aside the idea.

The legend of Newton and the apple satisfies a widely felt need to attribute the origin of theories to observations rather than to thinking. And if associating an apple with knowledge gains from evoking an earlier story involving Adam, Eve, a serpent, and an apple, so much the better. Another apple that supposedly changed history, also for the worse, was the golden apple thrown down in a fit of anger by the goddess of discord, and awarded by Paris, prince of Troy, to the most beautiful

(continued)

(Continued)

woman in the world—Aphrodite, who then promised him Menelaus's wife Helen, thus setting off the Trojan War. Newton's apple, in contrast, rendered great service to humankind.

Caricaturing the conception of the idea of universal gravitation as following inevitably from an apple falling by chance on any person's head rather than as the result of inspiration and years of subsequent thought and effort by a genius belittles Newton's accomplishment. It does, however, make for a delightful cartoon: three studious gentlemen sitting under separate trees somewhat apart, pencils and notepads in hand, waiting for ideas to appear; from one tree dangles a *huge* apple, somewhat larger than the scholar under it; one of the other men says, "Nothing yet . . . How about you, Newton?"

Sun and an impressed motion. Neither Wren nor Hooke, however, had the geometrical ability to show what orbit would result from an inverse square force of attraction: in effect, to derive Kepler's laws of planetary motion from principles of dynamics.

At a meeting of the Royal Society in January 1684 Wren, Hooke, and the astronomer Edmond Halley discussed the problem. Halley admitted that he had failed to demonstrate the laws of celestial motion from an inverse square force. Hooke claimed that he could do so. Wren doubted that Hooke could and offered a prize of a book worth 40 shillings to anyone who could produce a demonstration. Hooke replied that he intended to keep his solution secret until others failed, and thus would value his solution more.

In August, Halley visited Newton in Cambridge and while there asked him what he thought the curve would be that would be described by the planets, supposing the force of attraction toward the Sun was the reciprocal to the square of their distances from the Sun. Newton replied immediately that it would be an ellipse. Halley asked him how he knew this. Newton replied that he had calculated it. He could not, however, find the calculation among his papers. He promised to do the calculation again and to send it to Halley.

In November, Halley received from Newton a nine-page essay, *De motu corporum in gyrum* (*On the Motion of Bodies in an Orbit*). Halley recognized the importance of the work and reported it to the Royal Society in December.

Meanwhile, back in Cambridge, Newton was extending his geometrical demonstrations to additional phenomena and making the calculations more precise. He was completely absorbed in this work to the exclusion of virtually everything else. An acquaintance wrote that Newton often forgot to eat and sometimes walking in the garden would suddenly stop, turn about, and run up the stairs to his room, where he would begin writing while standing at his desk, not even taking time to draw up a chair to sit in.

The result of this frenzied concentration was one of the greatest and most influential and important books ever written: the *Principia*, or, to give the book its full title, *Philosophiae naturalis principia mathematica* (*Mathematical Principles of Natural Philosophy*). Again, Halley recognized the worth of Newton's work, and he communicated the manuscript to the Royal Society. Now clerk to the Society, Halley reported to its members that Newton "gave a demonstration of the Copernican hypothesis as proposed by Kepler, and made

The Moon's Orbit

If Newton actually did perform his Moon test calculation in 1666, he might have been inspired by an idea that only appears in print in A Treatise of the System of the World. This attempt to popularize Book III of the Principia, System of the World, came out in 1728, a year after Newton's death.

Figure 17.2: Shooting cannon balls on the Earth. From Isaac Newton, A Treatise of the System of the World. Image copyright History of Science Collections, University of Oklahoma Libraries.

Imagine a cannon on a mountain top V shooting cannon balls horizontally. As larger charges of powder are employed, the cannon balls go farther. The first lands at D, the second at E, the third at F. Increasingly larger charges of powder propel the next three cannon balls to B, to A, and finally to V, in effect orbiting the Earth.

Now think of the Moon's orbit as the result of the same force (gravity) causing cannon balls to fall to the Earth or remain in orbit around the Earth.

The Moon would move in a straight line (as if shot from a cannon) from M toward M' were it not pulled toward the Earth by the force of gravity. The Moon's resulting curved path bends below the horizontal line; the Moon falls from X to X'.

If the time from M to M' is 1 second, the distance the Moon is observed to fall from X to X' is 1.37 millimeters. A cannon ball near the Earth's surface falls about 5 meters (3,600 times more than 1.37 millimeters) in one second. Thus the force on the cannon ball is calculated to be 3,600 times as great as the force on the Moon.

The radius of the Earth is approximately 4,000 miles. The radius of the Moon's orbit is approximately 240,000 miles, 60 times the Earth's radius. Sixty times 60 is 3,600, which is the ratio of the forces on a cannon ball and on the Moon. Thus the force of gravitational attraction from the Earth pulling down a cannon ball and holding the Moon in its orbit decreases in proportion to the square of the distance.

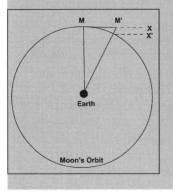

Figure 17.3: The Moon in orbit around the Earth. Gravity pulls the Moon down the distance between X and X' in the time the Moon otherwise would have moved horizontally from M to M'.

Or it would have been so calculated, had Newton had at hand the correct value for the radius of the Earth. But he was in the country, away from the university library. Supposedly he used an incorrect value in his calculation and thus concluded, temporarily and erroneously, that some other force must also be involved. Years later he would plug the correct value into the calculation and all would be right.

out the phenomena of the celestial motions by the supposition of a gravitation towards the center of the Sun decreasing as the squares of the distances therefrom reciprocally" (Westfall, *Never at Rest*, 444–45). Halley undertook to see the book published at his own expense when the Royal Society shirked its duty.

The title page reflects Newton's intent to refute Descartes' philosophical principles, which had appeared under the title *Principia philosophiae*. Descartes' title is incorporated in Newton's, and lest readers miss the allusion, set in boldface and larger type. In the third edition, the words *Philosophiae* and *Principia* were in scarlet color. Newton's philosophy was not the pretentiously new metaphysical philosophy of Descartes, but a *natural* philosophy founded securely on *mathematics;* in place of philosophical principles was a natural philosophy consisting of mathematical principles.

In his *Principia*, Newton "laid down the principles of philosophy; principles not philosophical but mathematical: such, namely, as we may build our reasonings upon in philosophical inquiries. These principles are the laws and conditions of certain motions, and powers or forces" (*Principia* III, Introduction). His four rules of reasoning were the following:

I. We are to admit no more causes of natural things than such as are both true and sufficient to explain their appearances.

II. Therefore to the same natural effects we must, as far as possible, assign the same causes.

III. The qualities of bodies, which admit neither intensification nor remission of degrees, and which are found to belong to all bodies within the reach of our experiments, are to be esteemed the universal qualities of all bodies whatsoever. (*Principia* III, Rules of Reasoning)

From this rule it followed that:

Lastly, if it universally appears, by experiments and astronomical observations, that all bodies about the Earth gravitate towards the Earth, and that in proportion to the quantity of matter which they severally contain; that the Moon likewise, according to the quantity of its matter, gravitates towards the Earth; that on the other hand, our sea gravitates towards the Moon; and all the planets one

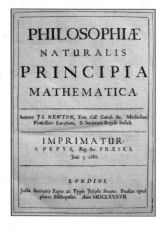

Figure 17.4: Title page from the *Principia*, 1687. Latin was still the scholarly language of Europe in Newton's time. The *c* in *Principia* is pronounced as if it were a *k*. The *Principia* was first published in 1687 in London in an edition of perhaps as many as 500 copies. Another 750 copies were printed at Cambridge in 1713, and 800 more (plus 25 on large paper and 10 on extra large paper) at London in 1726. Also during Newton's lifetime unauthorized editions appeared in Amsterdam in 1714 and 1723. Image copyright History of Science Collections, University of Oklahoma Libraries.

towards another; and the comets in like manner towards the Sun; we must, in consequence of this rule, universally allow that all bodies whatsoever are endowed with a principle of mutual gravitation. (*Principia* III, Rules of Reasoning)

Newton's grand generalization "that all bodies whatsoever are endowed with a principle of mutual gravitation" was based on a narrow range of evidence. He didn't know anything about distant stars, about tiny entities visible only in microscopes, about particles of light, or even very much about common bodies on the Earth that might be manipulated in ordinary laboratory settings to yield evidence of mutual gravitation. For example, in 1798 the British scientist Henry Cavendish would measure the attraction between two lead spheres with a torsion balance, and then from that value calculate the mass of the Earth (to within 1% of the currently accepted value). Still, what scientist has ever had all the possible facts in hand before leaping to a generalization? Uniformity is a necessary assumption of science. From a limited body of data scientists draw overly broad generalizations, which they then test and often subsequently refine by dividing a class of objects previously presumed universal into different types with different laws.

The fourth and final rule of reasoning followed:

> IV. In experimental philosophy we are to look upon propositions inferred by general induction from phenomena as accurately or very nearly true, notwithstanding any contrary hypotheses that may be imagined, till such time as other phenomena occur, by which they may either be made more accurate, or liable to exceptions. (*Principia* III, Rules of Reasoning)

Edmond Halley

Born into a prosperous London family engaged in business, Halley attended Oxford University. While a student there he made astronomical observations and reported his results in the *Philosophical Transactions of the Royal Society.* In 1676 Halley sailed to St. Helena, the southernmost territory then under English rule. At the request of King Charles II, the East India Company provided for Halley's transportation on one of their ships and for his maintenance on the island. His father paid for his instruments. Halley measured positions of stars and observed eclipses and a transit of Mercury across the face of the Sun. He dedicated his chart of stars of the Southern Hemisphere, with a newly depicted constellation, *Robur Carolinum (Charles' Oak)*, to Charles. Soon after Halley's return to England, the Royal Society elected him a fellow, at the early age of 22.

In 1684 Halley asked Isaac Newton what sort of planetary orbit an inverse square force would produce, a question that led to the *Principia*. For the second edition of the *Principia*, Halley in 1695 undertook to calculate comet orbits. He soon realized that the comets of 1531, 1607, and 1682 had similar orbits: it was the same comet revolving around the Sun in an elliptical orbit. Halley died before the predicted return of his comet in 1758. He also died before the predicted transits of Venus of 1761 and 1769, but he left detailed instructions for calculating the size of the solar system from their observation.

In the 1690s Halley discovered from his study of solar and lunar eclipse observations by an Arab astronomer in the ninth century A.D. that the Moon's mean motion had accelerated. And in 1718, Halley announced that a comparison of stellar positions in his day with those measured by Hipparchus revealed that in the course of 1,800 years several of the supposedly fixed stars had altered their places relative to other stars: they had motions of their own.

Halley also found time to undertake the attempted salvage of gold from a sunken ship, *(continued)*

(Continued)

which involved improving the diving bell; to survey harbors on the Dalmatian coast for potential use by the British fleet in the anticipated war of the Spanish succession; and to serve for several years as captain (an extremely rare appointment for a civilian) of his Majesty's ship *Paramore,* carrying out a study of tides in the English Channel and of magnetic variation in the Atlantic Ocean. During his second voyage, Halley discovered and took possession, in his Majesty's name, of a small uninhabited volcanic island in the South Atlantic.

Halley was appointed Savilian Professor of geometry at Oxford in 1704. There he produced editions of several classics of ancient Greek geometry. He was elected Secretary of the Royal Society in 1713 and resigned the position in 1720 when he succeeded John Flamsteed as Astronomer Royal, though he remained active in the Society. Halley carried out astronomical observations until a few months before his death, in 1742.

Not until the third edition, of 1726, did Newton distinguish propositions from hypotheses. He added: "This rule we must follow, that the argument of induction may not be evaded by hypotheses" (*Principia* III, Rules of Reasoning).

Following the argument of induction, Newton ostensibly began with phenomena. This was the strategy Kepler had adopted in his *Astronomia nova* in response to criticism for having begun with a priori theory in his earlier *Mysterium cosmographicum.* Inductivist thinking would become so strongly entrenched after the *Principia* that even Newton's disciples presumed that he had relied on Kepler's laws as empirical premises. Universal gravitation came to be seen (incorrectly) as an inductive achievement from the facts of planetary motion encapsulated in Kepler's laws.

The first phenomenon listed by Newton was that the radii of Jupiter's moons sweep out equal areas in equal times and the periodic times are as the $^3/_2$ power of their distances. "This we know from observations," Newton wrote (*Principia* III, Phenomenon I). Observations revealed the same phenomenon for Saturn's moons and for the five planets revolving about the Sun. Here Newton acknowledged Kepler as the first to have observed this proportion. With only one satellite then revolving around the Earth, Kepler's third law, involving ratios for two satellites, was not applicable; but the Moon was observed to sweep out equal areas in equal times.

Next in the *Principia* came propositions. The implication was that they were induced from the phenomena. For each phenomenon, Newton proposed that the forces by which the moons or planets are "continually drawn off from rectilinear motions, and retained in their proper orbits" tend to the central body (the Sun in the case of the planets; a planet in the case of its moons) and are "inversely as the squares of the distances of the places of those planets [or moons] from that center" (*Principia* III, Proposition I). Newton also proposed that the planets move in ellipses. In a letter to Halley Newton claimed credit for *establishing* this proposition. He believed that Tycho Brahe's data were not adequate to warrant or guarantee the hypothesis and that Kepler had merely guessed at the ellipse.

The argument of induction is that from observations or phenomena, theories or propositions follow. After enough observations are accumulated, somehow a theory will appear in the human mind. Does it really matter, though, where a theory comes from? From observation by induction or from out of the mystery of the human mind by some less rational and less objective process? Justification for a theory is not necessarily to be found in its origin.

Newton's great achievement was to show by deduction that from the theory of universal gravitation all the observed phenomena mathematically follow. Immediately after the propositions, he wrote: "Now that we know the principles on which they [the phenomena] depend, from these principles we deduce the motions of the heavens a priori" (*Principia* III, Proposition I). This Newton proceeded to do. Not only Kepler's laws, but many additional phenomena fell under Newton's geometrical onslaught.

From ancient observations, Halley had noticed that seemingly the same comet had appeared four times at intervals of 575 years, most recently in 1680. Newton now developed a geometrical procedure for determining from observations the orbit of a comet moving in a parabola. Then he compared observations of several comets with their predicted places computed from his theory. Newton concluded that "from these examples it is abundantly evident that the motions of comets are no less accurately represented by our theory than the motions of the planets commonly are by the theory of them; and, therefore, by means of this theory, we may enumerate the orbits of comets" (*Principia* III, Proposition XLII, Problem XXII).

The Moon, its motion around the Earth disturbed by the Sun's gravitational force, presented considerable difficulties. It moves faster and its orbit is less curved, and therefore the Moon approaches nearer to the Earth, in the syzygies (when the Moon lies in a straight line with the Earth and the Sun: at opposition or conjunction) than in the quadratures (at 90 degrees from the line joining the Sun and the Earth)—except when the eccentricity of the orbit affects this motion. The eccentricity is greatest when the apogee of the Moon (the point in its orbit farthest from the Earth) is in the syzygies, and least when the apogee is in the quadratures. Moreover, the apogee goes forward, and with an unequal motion: more swiftly forward in its syzygies and more slowly backwards in its quadratures, with a net yearly forward motion. Furthermore, the greatest latitude of the Moon (its angular distance above or below the plane of the Earth and the Sun) is greater in the quadratures than in the syzygies. Also, the mean motion of the Moon is slower when the Earth is at perihelion (nearest the Sun) than at aphelion (farthest from the Sun).

These were the principal inequalities in the motion of the Moon taken notice of by previous astronomers. In a remarkable mathematical tour de force, Newton demonstrated that all these inequalities followed from the principles he had laid down. There remained, however, "yet other inequalities not observed by former astronomers by which the motions of the Moon are so disturbed that to this day we have not been able to bring them under any certain rule" (*Principia* III, Proposition XXII, Theorem XVIII). Newton also demonstrated geometrically that the flux and reflux of the ocean tides arose from the actions of the Sun and Moon.

Newton's geometrical demonstrations were not restricted to celestial bodies. He also deduced that, other things being equal, the squares of the times of oscillation are as the lengths of pendulums; the weights are inversely as the squares of the times if the quantities of matter are equal; the quantities of matter are as the weights if the times are equal; and if the weights are equal, the quantities of matter are as the squares of the times.

Although a theory can never be absolutely proven because the world could have been made in some different manner that nonetheless has the same set of observational consequences, a theory does gain in confidence with each deduced phenomenon that is observed. Newton had deduced from an inverse square force of gravity an impressive number of effects.

There remained the question, what is gravity? Newton acknowledged that "hitherto we have explained the phenomena of the heavens and of our sea by the power of gravity, but have not yet assigned the cause of this power" (*Principia* III, General Scholium). Some things Newton did know:

> This is certain, that it [gravity] must proceed from a cause that penetrates to the very centres of the Sun and planets, without suffering the least diminution of its force; that operates not according to the quantity of the surfaces of the particles upon which it acts (as mechanical causes used to do), but according to the quantity of the solid matter which they contain, and propagates its virtue on all sides to immense distances, decreasing always as the inverse square of the distances. (*Principia* III, General Scholium)

But he didn't know what gravity was. In the second and third editions of the *Principia*, Newton wrote: "But hitherto I have not been able to discover the cause of those properties of gravity from phenomena, and I feign no hypotheses. . . ." (*Principia* III, General Scholium).

Writing in Latin, Newton's actual words were "hypotheses non fingo." Sometimes this phrase is translated as "I frame no hypotheses." But Newton meant *feign*, in the sense of pretend: to present hypotheses that could not be proved and were probably false. Elsewhere he framed hypotheses that could be demonstrated, as all scientists do.

Newton continued. Because in experimental philosophy or science the emphasis should be on observed phenomena, propositions obtained by induction from them, and predicted phenomena deduced from propositions, "whatever is not deduced from the phenomena is to be called an hypothesis; and hypotheses, whether metaphysical or physical, whether of occult qualities or mechanical, have no place in experimental philosophy. In this philosophy particular propositions are inferred from the phenomena, and afterwards rendered general by induction" (*Principia* III, General Scholium).

For justification of experimental philosophy, Newton directed his readers to its remarkable success: "Thus it was that the impenetrability, the mobility, and the impulsive force of bodies, and the laws of motion and of gravitation, were discovered" (*Principia* III, General Scholium). Newton argued for setting aside the question of what gravity is and instead be content with a mathematical description of its effects. He wrote: "And to us it is enough that gravity does really exist, and act according to the laws which we have explained, and abundantly serves to account for all the motions of the celestial bodies, and of our sea" (*Principia* III, General Scholium). Newton would have liked to explain the *cause* of gravity, but he wasn't able to, so he settled for demonstrating that the motions of the celestial bodies could be deduced mathematically from an inverse square force, whatever anyone might imagine the metaphysical or occult qualities of that force to be.

To explain motion Aristotle had attributed it to a final cause or purpose. Newton didn't explain motion; rather he demonstrated that the observed motions could be deduced mathematically from an inverse square force. One should abstain—as Aristotle had not—from speculating about the nature of that force. Molière's quack doctor might explain the sleep-inducing power of opium in terms of its dormative potency; Newton did not explain gravity.

Also abandoned rather than answered by Newton was the question that had dominated astronomy since Copernicus and was at the center of the Galileo controversy: was the Earth or the Sun at the center of the universe? For Newton, the important center was the center of gravity, and neither the Earth nor the Sun necessarily resided there. (Since most of the mass in our planetary system is contained in the Sun, the Sun very nearly coincides with the center of gravity of the solar system; but they would not coincide in a system with a different distribution of mass.)

Yet another matter completely lost sight of in Newton's new science was the ancient Aristotelian distinction between terrestrial and celestial matter and their physical laws. Newton blithely united the terrestrial and the celestial.

The abandonment of scientific beliefs, values, and worldviews and their replacement with incompatible or even incommensurable new paradigms is the essence of a scientific revolution. Some historians accept the replacement of one theory with a second, incompatible theory as a revolution. Others might withhold the word *revolution* for the replacement of one worldview with another worldview so incommensurable that rival proponents cannot agree on common procedures, goals, and measurements of success or failure.

The Ptolemaic and Copernican systems were incompatible but not incommensurable. While they predicted different results, such as the appearance of the phases of Venus, yet they were judged by the mutual standard of how well each saved the phenomena with systems of uniform circular motions.

Descartes' vortex system and Newton's gravity were more incommensurable than incompatible. They had different goals and different measurements of success or failure. Descartes had insisted on an explanation of the cause of gravity. Newton abandoned that quest and argued instead that it was enough that gravity acted according to an inverse-square force law and accounted for all the motions of the celestial bodies. Though Descartes' later followers eventually agreed that any successful vortex theory would have to account for Kepler's laws, Descartes himself had ignored Kepler's discoveries, if indeed he had known of them. Newton, on the other hand, asked that his theory be judged by its success in accounting for Kepler's laws. Descartes' and Newton's theories were not incompatible, at least initially, with a mutual standard against which they could be judged, but incommensurable: not comparable on any mutually agreed upon basis. Newton established a new worldview and a new way of doing science.

Even poets were impressed. Alexander Pope rhymed:

Nature, and Nature's laws lay hid At Night
God said, *Let Newton be!*, and All was *Light*. (Feingold, *Newtonian Moment*, 144)

18

THE NEWTONIAN
REVOLUTION

Revolutions, scientific and social, consist of more than an initial press release followed by a victory celebration. Often, they are drawn out wars between opposing forces, with the outcome long in doubt. Copernicus declared a heliocentric system in 1543, but more than half a century later Galileo was fighting valiantly; nor did he win every battle. Near the end of the seventeenth century Isaac Newton quantitatively reproduced Kepler's mathematical kinematics of the heavens with a new celestial dynamics based on an inverse-square force. It remained to be seen if Newton's gravity could win out over the French philosopher René Descartes' vortices, and if gravity did triumph, what repercussions the victory might have beyond astronomy.

A propaganda effort on Newton's behalf was already underway even before the *Principia* appeared in print. Edmond Halley wrote to a German professor of mathematics and physics to inform him of the new theory of universal gravitation brilliantly investigated by Newton. And Halley also wrote to the physician to the duke of Würtemburg to inform him of "a truly outstanding book" written by perhaps the greatest geometer ever to exist, which would prove "how far the human mind properly instructed can avail in seeking truth" (Feingold, *Newtonian Moment*, 31). Also before the *Principia* was published, there appeared in the *Philosophical Transactions of the Royal Society of London* (edited by Halley) an anonymous review (written by Halley) of Newton's incomparable treatise, which provided "a most notable instance of the extent of the powers of the Mind," and showed "what are the principles of Natural Philosophy, and so far derived from them their consequences" that Newton seemed "to have exhausted his Argument, and left little to be done by those that shall succeed him" (Feingold, *Newtonian Moment*, 31).

Actually, the concerted efforts of many mathematician-astronomers during much of the following century would be required to work out details left undone by Newton. To write the history of eighteenth-century astronomy is

Voltaire's *Lettres Anglaises*

Royal permission for publication of Voltaire's *Letters Concerning the English Nation* had been withheld, and enraged authorities issued a warrant for his arrest. Tipped off by one of the King's ministers, Voltaire had already fled to Madame du Châtelet's chateau, conveniently located, were he pursued, near the border with Lorraine, then an independent province. The couple enjoyed both a romantic and an intellectual relationship. Her husband tolerated the presence of Voltaire, who loaned the Marquis money to renovate the chateau.

Voltaire had left behind in Paris his publisher, who was thrown into the Bastille, and copies of his impious book implying that French civilization was deficient, which were burnt by the public hangman. Scandal and the distinction of being banned in Paris fueled sales of the book, which topped 20,000 within five years.

to write about the continuation of the research program, or paradigm, set by Newton.

In hindsight, victory was inevitable. For nearly a century, however, Newtonians battled a strong rival army. In the middle of the seventeenth century, before Newton's work, Descartes had proposed that the universe consists of huge whirlpools, or vortices, of cosmic matter. Our solar system was one of many whirlpools, its planets all moving in the same direction in the same plane around a luminous central body. Moons also were carried by vortices around their planets.

The French literary giant François-Marie Arouet de Voltaire wittily summed up the situation in his 1733 *Lettres philosophiques*, also known as *Lettres anglaises*, or *Letters Concerning the English Nation*. A Frenchman arriving in London would find philosophy, like everything else, very much changed there. He left the world full, and found it empty; he left the world a plenum, and found it a vacuum. In Paris the universe was seen composed of vortices of subtle matter; but nothing like vortices were to be seen in London. In Paris everything was explained by a pressure that nobody understood; in London everything was explained by an attraction that nobody understood either.

Publication in 1737 of Voltaire's *Elémens de la philosophie de Newton* (*Elements of Newton's Philosophy*), according to a reviewer, resulted in all Paris resounding with Newton, all Paris stammering Newton, and all Paris studying and learning Newton. Madame du Châtelet, who had assisted Voltaire with the *Elements* (the frontispiece depicts her held in the sky by winged cherubs, one breast bared, and holding a mirror reflecting rays of light coming from behind Newton—who is wearing a toga and is seated on a throne of clouds—to Voltaire, busy writing at his earthbound desk), also reviewed the *Elements*, favorably and anonymously. In a private letter, this Minerva (the goddess of reason) of France wrote that publication of Voltaire's books was necessary because the French were blissfully ignorant of flaws in Cartesianism evident to everyone else in Europe and thus were unable to participate in the progress which Newton's discoveries would make possible.

Voltaire likened himself to a man blind in one eye expostulating with men blind in both eyes. (Earlier, Voltaire had written that "Before Kepler all men were blind; Kepler was one-eyed, and Newton had two eyes" [Feingold, *Newtonian Moment*, 31]). Voltaire also likened Newton, the greatest man who had ever lived, to the kingdom of heaven, and the French to the little children in Matthew 19:14: "But Jesus said, Suffer little children, and forbid them not, to come unto me: for of such is the kingdom of heaven." Voltaire was instrumental in spreading the Newtonian revolution; as late as 1774 his books were still praised as the only source for gentlemen not learned in the sciences to turn to for information about Newton. That was in France. In England, as early as 1702, a few university students, if not cultured gentlemen and ladies, had access to Newton's science in two new textbooks: John Keill's *Introductio ad varem physicam* (*Introduction to the True Physics*) and David Gregory's *Astronomiæ, physicæ et geometriæ elementa* (*Elements of Astronomy, Physical and Geometrical*).

In the same year as the appearance of Voltaire's *Elements of Newton's Philosophy,* the Newtonian revolution gained yet more ground, from measurements of the shape of the Earth. Newton had predicted that the Earth was flattened at the poles; Cartesians said it must be flattened at the equator. As Voltaire put it: "In Paris you see the Earth shaped like a melon, in London it is flattened on two sides" (Feingold, *Newtonian Moment,* 100). Local measurements by Jacques Cassini, one of four generations of Cassinis to direct the Paris Observatory from its founding in 1669 until they were driven out in 1793 during the French Revolution, supported the Cartesians. Pierre-Louis Moreau de Maupertuis, a young French mathematician sent by the Académie des Sciences to Lapland to make more decisive measurements, returned in 1737 as the "flattener of Earth and of the Cassinis," so Voltaire

Measuring the Figure of the Earth

In 1735 the French Académie des Sciences dispatched an expedition to Peru to measure the length of a degree of arc at the equator. At the same time, Maupertuis, a member of the Académie (and a teacher and probably lover, too, of Madame du Châtelet), proposed an expedition to the Arctic Circle to make measurements there to check and supplement those from Peru. Maupertuis was chosen to lead the expedition, accompanied by two astronomers from the Academy, a clock maker, a draftsman, a secretary, and a priest, plus the Swedish astronomer Anders Celsius.

Maupertuis's group of scientists left for Lapland in April 1736. A very rough three weeks of sailing delivered them from Dunkirk to Stockholm. They had planned to sail on to Tornio (now in Finland; then a province of Sweden) at the northern end of the Gulf of Bothnia, but a near mutiny within his party persuaded Maupertuis to change plans and make a longer trip on solid ground. The intrepid scientists set out in two purchased coaches on June 6 and arrived in Tornio on June 21 (a bus now leaves Stockholm at 6 in the evening and arrives at Tornio at 8:45 the next morning). The expedition's servants, instruments, and provisions had gone by boat and were waiting in Tornio.

Maupertuis had planned to lay out his degree of latitude for measuring over ice in the Gulf that winter, but the local people, while not sure that the Gulf would freeze over, were sure that south winds coming up without warning would open up any ice. Also, the islands in the Gulf were heavily wooded and not high enough to make measurements from. Fortunately, the Tornio River runs nearly north–south, and its valley has high peaks surrounding it, from which measurements could be made. The farthest of seven mountain peaks was 70 miles north of Tornio.

The mountains were rough and steep, but no worse than the swamps and thick brush elsewhere or the heat and the ferocious flies in

(Continued)

(Continued)

midsummer. Nor was the bitter cold of winter much relief. Some of their thermometers froze, encouraging Celsius later to develop the mercury thermometer. (He is best known for establishing the thermometer scale named after him: zero for the freezing point of water and 100 for boiling—actually, zero for boiling and 100 for freezing, but the scale was reversed after his death). Maupertuis was too busy to reply to Madame du Châtelet's letters, though he did find time to form an attachment with a Laplander so binding that she and her sister would accompany him back to France.

Its work finished, the expedition left Tornio in June 1737, this time by ship. They were shipwrecked in the Gulf but managed to save their precious records. They finally arrived back in Paris in September and made their report to the Academy in November. Madame du Châtelet called her hero Sir Isaac Maupertuis.

Meanwhile, the expedition to the equator continued, with one member dead of fever, a second gone insane, and a third killed by a mob. In England it was reported that "An unlucky accident has happened to the french Mathematicians in Peru. It seems they were shewing some french gallantry to the natives wives, who have murdered their servants destroyed their Instruments & burnt their papers, the Gentlemen escaping narrowly themselves. What an ugly Article will this make in a journal" (Jones, *Figure of the Earth*, 113). The international War of the Austrian Succession and the little local War of the Pyramids in Quito further disrupted the expedition. Their measurements were not completed until 1743, and the leader of the expedition, deciding to cross the Andes Mountains and follow the Amazon River to the Atlantic, did not make it back to Paris until 1745, 10 years after he had departed.

quipped. Maupertuis, however, gained little glory confirming to his compatriots a discovery predicted fifty years earlier by an Englishman (Feingold, *Newtonian Moment*, 106).

Years would pass before Voltaire's words and scientific measurements swept aside Descartes' natural philosophy, especially in France. Newtonianism's rival did not slink away quickly or quietly.

According to Descartes, natural phenomena were to be explained in terms of the mobility of particles of matter governed by a few general principles brought to bear on both terrestrial and celestial phenomena. These included the conservation of rest and motion as states of matter, and the primacy of rectilinear (straight line) motion. All change in motion was the result of percussion of bodies; one object could act on another only by contact. Descartes banned from scientific investigation occult phenomena, or causes hidden from the senses. Gravity was the result of celestial matter circulating about the Earth pushing all terrestrial matter toward the Earth. Descartes' mechanical, mechanistic cosmology was highly acceptable within the general seventeenth-century conception of the world as a machine. His explanations, though, were only qualitative redescriptions of phenomena in mechanistic terms.

Vortex theory enjoyed a major advantage over gravitational theory. Newton's mysterious attraction at a distance seemed a reversion to earlier, Aristotelian modes of thinking—a discredited way of thinking that the mechanical philosophy was intended to replace. Joseph Saurin, a member of the Académie des Sciences in France and a champion of Cartesian vortices, warned true philosophers against Newton's gravity, which "revived ideas like attraction and occult qualities that have rightly been decried" (Feingold, *Newtonian Moment*, 60). If they abandoned clear mechanical principles, "all the light we can hope to achieve is extinguished, and we will once again be returned to the ancient darkness of Peripatism [Aristotelianism] from which heaven wants to save us" (Feingold, *Newtonian Moment*, 60).

It was not easy to believe that the great masses of the planets were suspended in empty space and that they retained their orbits by an invisible influence. Even Newton admitted that the notion of one body acting on another at a distance without the mediation of anything else to convey the action was physically absurd. One of the great changes in scientific thought following from Newton's work would be the eventual abandonment of any necessity for a material aether pervading space to transmit gravitational forces, or light, or anything else. Not until 1887 would the famous Michelson-Morley experiment rule out any luminiferous light-bearing medium in the universe.

Newton, though bemoaning the lack of a material explanation for gravity, at the same time attacked Descartes' cosmic aether. Newton argued that its resistance would soon halt the planets' motions. Furthermore, Newton purported to show that the relation between tangential velocity and distance from the center in a vortex differed from Kepler's third law.

Neither Descartes nor his early followers had made any attempt to account for Kepler's observational results. Descartes may not have known of them, and his followers followed him uncritically. In fairness, it should be noted that no one prior to Newton regarded Kepler's laws as facts requiring explanation by physical theory.

The scientific struggle to bring vortex theory into agreement with observation began with Gottfried Leibniz, a German mathematician and philosopher, and a rival of Newton in the invention of the calculus, each accusing the other of plagiarism. Leibniz recognized Kepler's laws but also believed that some kind of fluid vortex was essential. A basic problem for Cartesians was that vortices rotated in circular layers, but the planets moved in elliptical orbits with varying distances from the central body. A vortex somehow had to maintain a planet's motion in a circular arc and also move the planet radially, toward and away from its central sun.

Leibniz argued that gravitational attraction was the result of a very tenuous matter, a second vortex, revolving in all directions around the Earth. This second vortex was independent of the vortex consisting of a coarser aether that carried the Earth around the Sun. Leibniz largely ignored the objection that comets, often inclined at large angles to the plane of circulation of the vortex and hence presumably largely beyond the control of the vortex, nevertheless obey Kepler's laws.

Through the first half of the eighteenth century, philosophers faced a choice between Descartes' vortices, agreeable to the intellect but overwhelmed by technical difficulties, and Newton's force of gravity, without physical explanation but increasingly in agreement with observed planetary motions.

Real bodies, it was generally agreed, were composed of particles separated by celestial matter. For Newton, the medium conveying action had to be immaterial, offering no resistance to bodies but moving them. Early in the eighteenth century, Saurin argued that if the particles of celestial matter were small enough to pass freely between the particles of dense bodies, the force of impact of the aether would be greatly reduced on dense bodies. Defenders of Descartes' vortex theory, however, generally insisted on a dense fluid.

There was considerable interest in extending vortex theory mathematically to reproduce in quantitative detail observed planetary motions. The Swiss mathematician Johann Bernoulli even won a prize from the Académie des Sciences in 1730 for his essay on the causes of elliptical orbits in a Cartesian vortex. Bernoulli hesitated to condemn Descartes' system, based on the clear and intelligible principle of celestial vortices, and substitute for it Newton's principle of gravity, for which no idea could be formed. Instead, Bernoulli attempted to assimilate Newtonian attraction and void with Descartes' vortices to yield a plausible physical explanation. A difference in density of the layers of the vortex would allow a planet to oscillate radially about an equilibrium position in which the densities of the layer and the planet were the same. The planet was still carried around the Sun by the circulation of the vortex, but the net composite motion could be an elliptical orbit; and Bernoulli claimed to have shown that the effects of vortices were compatible with Kepler's laws. In fact, he only outlined a solution; he failed to carry out the necessary mathematical demonstration. This task he excused himself from on the grounds that it would be too long and laborious. Indeed, a solution to the problem of vortex motion was beyond the best mathematical talent of the age.

A mere suggestion of the cause of elliptical orbits may have sufficed for the purpose of Bernoulli's 1730 essay, but the Cartesian program, producing qualitative but not quantitative explanation, was increasingly found insufficient. In 1740 Maupertuis, then the foremost French proponent of Newton, publicly declared his allegiance to Newtonian cosmology because of the failure of Cartesians to reconcile vortex theory with Kepler's mathematical laws of planetary motion.

Throughout the eighteenth century, Newtonians piled quantitative success upon quantitative success, though not in the form Newton himself had used. Geometry, long the accepted medium of mathematical proof, was so used by Newton, but those who came after him proved alternative mathematical procedures more productive. This change may explain at least partially why Newton's loyal followers in England made little progress compared to scientists on the continent less tightly tied to geometrical proofs and more ready to take up algebraic methods (developed by Descartes).

The solar system contains many planets, and the calculation of any one planetary orbit is not simply a matter of the gravitational attraction between that planet and the Sun. There are smaller but not negligible perturbational effects from the attraction of other bodies on any particular planet. For example, Jupiter and Saturn modify the motions of each other about the Sun. Also, the Sun alters the Moon's motion around Earth.

The Swiss mathematician Leonhard Euler helped develop necessary mathematical techniques to compute perturbation effects. First, he applied them to the Moon, and then, in 1748, to Jupiter and Saturn. The French

Académie des Sciences' prize topic for 1748 called for an explanation of the inequalities of motion that Jupiter and Saturn appeared to cause in each other's motions, their observed motions having been found inexplicable solely on the basis of Kepler's laws and an inverse-square force of gravity. Euler extended his analysis of the interaction of Jupiter and Saturn in another prize essay, in 1752. Joseph Lagrange and Pierre-Simon Laplace, both members of the Académie des Sciences, achieved further mathematical triumphs.

Perturbations of planets and satellites, the motions of comets, the shape of the Earth, precession (a slow conical motion of the Earth's axis of rotation caused primarily by the gravitational pull of the Sun and the Moon on the Earth's equatorial bulge), and nutation (a smaller wobble superimposed on the precessional motion of the Earth's axis) were all grist for the eighteenth-century's mathematical mill. In 1785 and 1787, Laplace seemingly resolved (many years later one of his solutions would be found to be only partially correct) the last major unexplained anomalies in the solar system: a large anomaly in the motions of Jupiter and Saturn and an acceleration of the Moon's orbital speed around the Earth.

Laplace's Nebular Hypothesis

Laplace's nebular hypothesis appeared in six versions between 1796 and 1835, the last after his death in 1827. The theory postulated that the Sun had once been a hot fluid extending beyond the present orbits of the planets. As the fluid cooled and condensed, gradually shrinking to the present size of the Sun, zones of material were left rotating around the Sun. These rings of gas shed by the contracting Sun condensed to form the planets. Similar processes centered on the planets produced their satellites.

Laplace's nebular hypothesis survived until the beginning of the twentieth century. Then the American geologist Thomas Chrowder Chamberlin argued persuasively against a molten Earth condensing from a gaseous mass on the grounds that: (1) all water vapor would have evaporated and been lost into space; and (2) that the observed distribution of mass and momentum, with most of the mass near the center of the solar system but most of the momentum in the outer planets, was an unlikely occurrence within Laplace's nebular hypothesis. It could, however, be accounted for if the solar system had been formed in a collision between a small nebula with large momentum and the periphery of a large nebula with very little momentum.

Laplace next took up the question of the origin of the solar system. Reflecting the new atheistic approach to nature of some of the scientists of the French Enlightenment, Laplace attempted to replace the hypothesis of God's rule with a purely physical theory that could also explain the observed order of the universe, particularly the remarkable arrangement of the solar system. He was successful, at least in his own mind. According to legend, when Napoleon asked Laplace whether he had left any place for the Creator, Laplace replied that he had no need of such a hypothesis.

Replacement of God's rule with a purely physical theory began the separation of science and religion, which previously were joined in Western thought. Theology had reigned as king of the disciplines, autonomous, the supreme principle by which all else was understood, its fundamental postulates and principles derived from divine revelation, interpreted and formulated within the tradition, and producing knowledge of ultimate value. Science had been

Intelligent Design: An Astronomical Perspective

Arguments for intelligent design often are centered in biology and directed against evolution. Nothing, however, better exemplifies the concept of intelligent design than Newton's words in the 1713 edition of his *Principia*, that "this most beautiful system of the Sun, planets and comets, could only proceed from the counsel and dominion of an intelligent and powerful Being" (*Principia* III, General Scholium). It seemed unlikely to Newton that random chance could have been responsible for the formation of the solar system.

The concerted efforts of many mathematician-astronomers during much of the eighteenth century were required to work out details left undone by Newton and to remove all need for God's intervention. According to legend, when Napoleon asked Laplace whether he had left any place for the Creator, Laplace replied that he had no need of such a hypothesis.

The concept of intelligent design, banished by Laplace in the case of the solar system, has come up again for the universe as a whole. Philosophers, contemplating the remarkable set of physical coincidences that seems to have been necessary for life as we know it to exist, have come up with an equally remarkable label for their thoughts: the *anthropic cosmological principle*. In its weak form, the principle states that the universe must be such as to admit and sustain life. From Descartes' "I think, therefore I am," we proceed to "I am, therefore the nature of the universe permits me to be." In its strong form, the anthropic cosmological principle states that the universe was created and fine-tuned so that intelligent life could evolve in it. This is the philosophical equivalent in cosmology of intelligent design in biology.

Repeatedly throughout history, complexities so remarkable that initially they were attributed to God eventually have been explained by plausible physical mechanisms. Already, the new inflationary universe theory has explained much of the remarkable set of coincidences that led to formulation of the anthropic cosmological principle.

Linking religious belief to any temporary stage of advancing scientific knowledge is likely to prove hazardous to religion. As early as the fourth century A.D., Saint Augustine counseled that no scientific doctrine should ever be made an article of faith, lest some better-informed heretic might exploit misguided adherence to the doctrine to impugn the credibility of proper articles of faith. In Galileo's time, a church official remarked that the Bible tells us how to go to heaven, not how the heavens go. Galileo, himself, cited Augustnine's advice in urging the Church not to condemn Copernican theory, especially while new evidence from the telescope was still coming in.

Only an unfortunate concatenation of circumstances culminated in the clash between Galileo and Catholic authorities. It was far from inevitable. Aristotelian philosophers in Italian universities succeeded in bringing the Church into battle on their side against Galileo, whose science, he realized, was "in contradiction to the physical notions commonly held among academic philosophers" and "stirred up against me no small number of professors" (Drake, *Discoveries*, 176). They had "resolved to fabricate a shield for their fallacies out of the mantle of pretended religion and the authority of the Bible" (Drake, *Discoveries*, 176).

Theologians understand this historical lesson, and the current battle over intelligent design is not between science and religion. Rather, the real battle, as it was in Galileo's time, is between questioning scientists seeking and testing naturalistic explanations of phenomena and a few opponents seeking to entangle unwitting allies, particularly school boards this time, in an effort to replace inquiry and reason with superstition.

merely a handmaiden, neither controlling fundamental knowledge nor ways of getting at it, its truths holding a lower logical status and value. Now, flush with triumphant reductions of all known phenomena of the solar system to the universal law of gravity, science began to displace religion as the source to which people turned for inspiration, direction, and criteria of truth. In the future, religion and politics increasingly would appeal to science for legitimacy.

Kepler's cosmology was strongly Christian. He was convinced that the Creator had used mathematical archetypes to design the universe. This religious belief drove Kepler's research and shaped his results. Newton, too, had been convinced that he was exploring and demonstrating God's wonders. It seemed unlikely that random chance could have been responsible for the formation of the solar system, and Newton wrote in the 1713 edition of his *Principia* that "this most beautiful system of the Sun, planets and comets, could only proceed from the counsel and dominion of an intelligent and powerful Being" (*Principia* III, General Scholium). When it was realized that irregularities in planetary motions caused by the disturbing influence of other planets would increase until the system wanted reformation, Newton took the necessity of God's occasional reformation as further evidence of his existence. On Newton's marble tomb in Westminster Abbey (he was the first scientist to be buried there) are the words "he vindicated by his philosophy the majesty of God mighty and good . . . " (Feingold, *Newtonian Moment*, 170).

Theological implications of Newton's cosmology were criticized in 1715 in a letter from Leibniz to Caroline, Princess of Wales. Newton's friend Samuel Clarke answered in a letter to Caroline, which she forwarded to Leibniz. In the course of the debate, Leibniz wrote five letters, and Clarke five replies, all of which were published, in 1717.

Leibniz charged that Newtonian views were contributing to a decline of natural religion in England. The implication that God occasionally intervened in the universe, much as a watchmaker has to wind up and mend his work, derogated from his perfection.

Clarke admitted that God had to intervene in the universe, but only because intervention was part of his plan. Indeed, the necessity of God's occasional reformation was, for Newtonians, proof of God's existence.

Eventually, Laplace demonstrated that the gravitational interactions of Jupiter and Saturn were self-correcting, not in need of divine intervention, and he also proposed a mechanism for the formation of the solar system. Replacement of the necessity of God in the universe with plausible physical hypotheses may have been inevitable as the Newtonian revolution played out, however convinced had been the revolution's founding fathers that they were exploring and demonstrating God's wonders.

Great as the Newtonian triumph was in science, it had an even greater effect on other realms of thought and on the history of Western civilization. Discovering more of the essential core of human knowledge than anyone before him, Newton illustrated the amazing ability of human reason, and through his success encouraged others to apply reason to other subjects. The eighteenth century became the century of enlightenment, in which critical human reason freed people from ignorance, from prejudices, and from unexamined authority, including religion and the state.

Newton, Astrotheology, and Galaxies

Eighteenth-century belief in the orderliness of the universe made determination of that order an important theological, philosophical, and scientific endeavor for astrotheologians. William Whiston, Newton's successor in the Lucasian Chair of Mathematics at Cambridge University, from 1702 to 1710, when he was charged with heresy and dismissed from the university, argued that the system of the stars, the work of the Creator, had a beautiful proportion, even if frail man were ignorant of the order. And William Derham, an ordained priest in the Church of England, a vicar and royal chaplain, expressed a similar belief in his 1715 *Astrotheology; or, a Demonstration of the Being and Attributes of God, from a Survey of the Heavens.*

Newtonian gravitational theory practically demanded a continual miracle to prevent the Sun and the fixed stars from being pulled together. If the stars were moving in orbits around the center of a system, however, like the planets move around the Sun, the result could be a stable system rather than gravitational collapse.

In 1718 Halley reported that three bright stars were no longer in the positions determined by ancient observations. The weight of tradition was so heavy, however, that even Halley continued to call the stars "fixt stars."

Eventually Halley's discovery suggested to the self-taught English astronomer Thomas Wright that the stars might be revolving around their center of gravity, and thus prevented from falling into their center, just as the planets are prevented from falling into the Sun. In 1750 Wright proposed a model for the Milky Way, a luminous band of light observed circling the heavens.

The German philosopher Immanuel Kant was inspired by an incorrect summary of Wright's book and by the paradigm of the Newtonian solar system. Kant's explicitly expressed intent was to extend Newtonian philosophy; the subtitle of his *Universal Natural History and Theory of the Heavens* was *An Essay on the Constitution and Mechanical Origin of the Whole Universe Treated According to Newton's Principles.*

Kant explained the Milky Way as a disk-shaped system seen from Earth, which was located in the plane of the disk. The arrangement of the stars, thought Kant, might be similar to that of the planets. Furthermore, the Newtonian system provided by analogy a physical explanation for a disk structure. Kant reasoned that the same cause that gave the planets their centrifugal force and directed their orbits to a plane could also have given the power of revolving to the stars and brought their orbits into a plane. Thoroughly imbued with a belief in the order and beauty of God's work, Kant went on to suggest that nebulous patches of light in the Heavens are composed of stars and are other Milky Ways, or island universes (galaxies similar to and beyond the boundaries of our own galaxy).

Between Whiston and Kant occurred a major change in worldview. At the beginning of the eighteenth century, the stable structure of the universe as well as its initial creation was believed the work of God. As the century progressed, God's role diminished. By the end of the century, laws of nature and sequences of mechanical events explained evolution from an initial chaos to the then-observed cosmological structure and also implied continuing change.

Political thinkers now had confidence that they could determine the natural laws governing human association, and the American and French revolutions followed. Albeit more a disciple of Descartes than of Newton, Charles-Louis de Secondat, Baron de Montesquieu, a French philosopher and writer, initiated attempts to apply the spirit of scientific inquiry to the study of man in society. In his 1748 *De l'esprit des lois* (*The Spirit of Laws*), Montesquieu noted similarities between scientific laws and social laws and hoped to find a few fundamental principles of politics. Thomas Jefferson, James Madison, and Alexander Hamilton were all enthusiastic readers of Montesquieu.

Newton's influence also pressed directly on early American statesmen, without the mediation of French political writers. Jefferson, who owned a portrait of Newton and a copy of his death mask, one of fewer than ten made, invoked "Laws of Nature" in the American Declaration of Independence. It was Jefferson's opinion that "we might as well say that the Newtonian system of philosophy is a part of the common law, as that the Christian religion is" (Feingold, *Newtonian Moment,* 160). On completing his second term as president of the United States in 1809, Jefferson said he was looking forward to giving up newspapers and reading instead Tacitus, Thucydides, Newton, and Euclid.

The British philosopher John Locke, who opposed authority and advocated the employment of reason in search of truth, also influenced American revolutionaries. Undoubtedly, Locke would have owed much to Newton, had the opportunity for debt arisen. But Locke's *Essay Concerning Human Understanding* was completed before the *Principia,* even if it wasn't published until 1690, three years after the *Principia.* In the second edition of his essay, Locke did praise Newton's success in applying many of the same principles of reasoning as Locke had used. He didn't mind hitching his star to Newton's, and historians subsequently have linked in their own studies the images of Newton and Locke in the Age of Reason.

Also enthusiastic about Newton was Benjamin Franklin, who so idolized Newton that he had his own portrait painted sitting at a desk inspired by a large bust of Newton towering over him. It was not uncommon to include a bust of Newton in commissioned portraits or a book in the background with Newton's name on the spine. Franklin, famous for his experiments on electricity, had more justification than most to link himself intellectually with Newton. The lieutenant governor of New York, a friend of Franklin and also a scientist of sorts, commemorated his purported explanation of the cause of gravity with a portrait of himself with his own book prominently at hand and a copy of Newton's *Principia* on a shelf behind him. The president of Yale College made no pretense to being a scientist, yet he too had a copy of the *Principia* included in his portrait, plus an emblem with comets moving in long ellipses. Newton's science had become an iconic image in art.

Back in the realm of politics, the British political philosopher Edmund Burke, who believed that even human taste had universal fixed principles and invariable and certain laws, had been sympathetic toward the American colonists and insisted that reason should govern political decisions. But he became dismayed by the violence and chaos in France, which he blamed on the undiscriminate application of the rules of reason. No more for Burke the cold abstract expression of an eternal logical system.

The strongest and most extensive link to Newton is found in the Scottish economist Adam Smith, who attempted to discover general laws of economics.

His cultural and intellectual milieu was dominated by the image of Newton, and he especially understood and appreciated Newton's achievement. While a youth, Smith wrote an essay, *The Principles which lead and direct Philosophical Enquiries; illustrated by the History of Astronomy.* Before reaching Newton in his history, Smith thought so-called scientific hypotheses were but arbitrary creations that might lend some coherence to the appearance of nature "by representing the invisible chains which bind together all these disjointed objects" (Hetherington, *Smith*, 281), but he doubted the reality of such associations. After contemplating Newton's achievement, however, Smith was persuaded that there are real connecting principles or chains that bind nature. He wrote:

> The superior genius and sagacity of Sir Isaac Newton therefore, made the most happy, and . . . greatest and most admirable improvement that was ever made in philosophy, when he discovered, that he could join together the movements of the Planets by so familiar a Principle . . . Such is the system of Sir Isaac Newton, a system whose parts are all more strictly connected than those of any other philosophical hypothesis. Allow his principle, the universality of gravity, and that it decreases as the squares of the distances increase, and all the appearances, which he joins together by it, necessarily follow. . . . His principles, it must be acknowledged, have a degree of firmness and solidity that we should in vain look for in any other system. The most sceptical cannot avoid feeling this. . . . And even we [Smith], while we have been endeavouring to present all philosophical systems as mere inventions of the imagination, to connect together otherwise disjointed and discordant phaenomena of nature, have insensibly been drawn in, to make use of language expressing the connecting principles of this one, as if they were the real chains which Nature makes use of to bind together her several operations. Can we wonder then, that it should have gained the general and complete approbation of mankind, and that it should now be considered, not as an attempt to connect in the imagination the phaenomena of the Heavens, but as the greatest discovery that ever was made by man, the discovery of an immense chain of the most important and sublime truths, all closely connected together by one capital fact, of the reality of which we have daily experience. (Hetherington, *Smith*, 281–82)

In his 1776 *Wealth of Nations*, Smith presented real connecting principles of economics. Copying Newton's strategy, Smith first listed phenomena, next obtained general laws—ostensibly by induction—and then deduced both the listed phenomena and further phenomena from the general laws. Newton's and Smith's answers to the question of the ultimate nature of their general principles are also similar. Unable to determine whether the propensity to exchange one thing for another was an original principle of human nature or a necessary consequence of the faculties of reason and speech, Smith argued that it was enough to know that the principle was universal, or common to all men. He would not frame or feign an uncertain hypothesis. Furthermore, Smith invoked Newton's principle of gravity both metaphorically and descriptively: "The natural price, therefore, is, as it were, the central price, to which the prices of all commodities are continually gravitating" (Hetherington, *Smith*, 284). The reference to gravity was no accident; two pages later, Smith repeated the phrase. And lest his readers fail to see the connection to Newton, Smith coyly called attention to his wording: "But though the market price of every particular commodity is in this manner continually gravitating, if one may say so, towards the natural price" (Hetherington, *Smith*, 284).

French economists also appreciated Newton. From their belief that government policy should not interfere with the operation of natural laws of economics came the phrase *laissez-faire, laissez-passer* ("allow to do, let things pass"). Charles Fourier, a French utopian socialist, despised laissez-faire liberalism and attributed his own discovery of a principle of human motivation to his musings on the fact that an apple in a restaurant cost fourteen sous, the same price as for a hundred apples in the country. To Adam's, Paris's, and Newton's historically significant apples, Fourier added his own. Newton had discovered the physical laws of universal attraction; Fourier, the laws of passional attraction, in opposition to morality, which taught people to war within themselves, resist their passions, and deny that God organized our souls.

Newton's influence also spread to the arts, especially music. The ancient Greeks had sought mathematical bases for music, and music was part of the quadrivium taught in medieval universities, along with astronomy, arithmetic, and geometry. The French composer and theorist Jean-Philippe Rameau, who composed several operas for which Voltaire supplied the librettos, believed that music was a science which should have definite rules drawn from evident principles known to us with the aid of mathematics. Rameau wanted to raise the status of music to the high level newly enjoyed by Newtonian science. In his 1722 *Traité de l'harmonie Reduite à ses Principes naturels* (*Treatise on Harmony reduced to its Natural Principles*), Rameau formulated principles of harmonic composition. When a better writer extracted the substance from Rameau's book and expounded it with lucidity and elegance in 1754, Rameau acquired the title of the "Newton of harmony," even if his theories owed more to his keen observation of music than to his employment of mathematics, or indeed any scientific method.

Literature, too, was affected by Newton's scientific achievement. Voltaire lamented, wittily as ever:

Fine literature is hardly any longer the fashion in Paris. Everyone works at geometry and physics. Everyone has a hand at an argument. Sentiment, imagination, and the finer arts are banished. A man who had lived under Louis XIV, if he came back, would not recognize the French; he'd think that Germans had conquered this country. Literature as such is perishing visibly. Not that I am angry that science is being cultivated, but I don't want it to become a tyrant that excludes all the arts. In France it is nothing but a fashion which succeeds another fashion and which will pass in its turn; yet no art, and no science, should simply be "in fashion." They should hold each other by the hand; we ought to cultivate them all the time. (Buchdahl, *Image of Newton*, 63–64)

Voltaire did not "see why the study of physics should crush the flowers of poetry. Is truth such a poor thing that it is unable to tolerate beauty?" (Buchdahl, *Image of Newton*, 64).

Nonetheless, science and the arts were beginning to drift apart. Poets were uncomfortable with what they saw as excessive rationalism and rigid determinism of nature's laws. They wanted to speak to the heart and to the imagination. Still, even Romantics in rebellion against Newtonianism recognized Newton's great achievement. Goethe was prouder of understanding optics and the science of colors better than Newton had than he was of his own literary achievements, which he acknowledged were equaled by other poets in his time and would be surpassed in the future. The German dramatist and poet Friedrich van Schiller wrote, in addition to his "Ode to Joy" sung in Beethoven's Ninth Symphony, that "Man had to be an animal before he knew that he was a spirit; he had to crawl in the dust before he ventured on the Newtonian flight through the universe" (Feingold, *Newtonian Moment*, 177). The concept of genius, previously reserved for artists and poets, was expanded by them to include Newton.

Samuel Johnson, the leading literary scholar and critic of the eighteenth century, wrote that Newton stood alone because he had left the rest of mankind behind him. Johnson might better have written that Newton had beaten a new path, one that the rest of mankind was now following. Jean D'Alembert, scientist and co-editor of the French *Encyclopedie,* which was intended to unify and popularize achievements of the new science, wrote:

> The true system of the world has been recognized . . . natural philosophy has been revolutionized . . . the discovery and application of a new method of philosophizing, the kind of enthusiasm which accompanies discoveries, a certain exaltation of ideas which the spectacle of the universe produces in us; all these causes have brought about a lively fermentation of minds. Spreading throughout nature in all directions, this fermentation has swept everything before it which stood in its way with a sort of violence, like a river which has burst its dams . . . thus, from the principles of the secular sciences to the foundations of religious revelation, from metaphysics to matters of taste, from music to morals, from the scholastic disputes of theologians to matters of commerce, from natural laws to the arbitrary laws of nations . . . everything has been discussed, analyzed, or at least mentioned. The fruit or sequel of this general effervescence of minds has been to cast new light on some matters and new shadows on others, just as the ebb and flow of the tides leaves some things on the shore and washes others away. (Buchdahl, *Image of Newton*, 62–63)

His new science linked through interrelationships with human values, culture, religion, society, and history in general, Newton washed away perhaps more ancient thought and superstition than any other individual in the tides of human affairs, and washed in more new and revolutionary thinking. His influence on so many aspects of eighteenth-century life and the continuing history of Western civilization is incredibly immense.

TIMELINE

3000 to 2000 B.C.	Sumerians control the lower part of the Tigris-Euphrates Valley.
1792 to 1750	Amorites ruled by Hammurabi, famous for his code of laws.
9th century	Beginning of the Assyrian Empire.
612	Destruction of Nineveh and the Assyrian Empire.
May 28 of 584	A solar eclipse, which modern science informs us took place on this date in what is now northern Turkey, brings an abrupt end to a battle between the Medes and the Lydians.
539	Persians capture Babylon.
480	Persians under Xeres I capture and burn Athens.
469–399	Socrates. Teacher of Plato.
431–404	Peloponnesian Wars between Athens and Sparta.
427–348	Plato. His book the *Republic* contains the Allegory of the Cave regarding the nature of reality, and a discussion of how to do astronomy. Simplicius of Athens in his commentary on Aristotle's book *On the Heavens* wrote around 500 that Plato had set as a task for astronomers to explain the apparently irregular motions of the planets, the Sun, and the Moon as a combination of circular motions with constant speeds of rotation.
ca. 408–347	Eudoxus. Student of Plato. Tried to save the phenomena of the planetary motions with a system of Earth-centered spheres rotating with constant speeds.

384–322	Aristotle. Author of the *Physics, On the Heavens,* and other books. His physics dominated western thought to the time of Copernicus. He established the use of demonstrative syllogisms, in which certain premises are stated, from which follow by deduction necessary conclusions.
380	Plato founds the Academy in Athens.
ca. 370	Callippus. Student of Eudoxus. Extended, with Aristotle's help, Eudoxus' system of concentric spheres in an attempt to save the phenomena of the planets.
330	Alexander the Great establishes the port city of Alexandria on the western edge of Egypt's Nile River Delta.
	Alexander the Great conquers the Persian Empire.
323	Alexander the Great dies.
ca. 310–230	Aristarchus of Samos. This "ancient Copernicus" propounded a heliocentric theory.
ca. 290	Ptolemy I founds the Museum in Alexandria, home to a hundred scholars subsidized by the government doing research and giving lectures, and where specimens of plants and animals were collected for study.
ca. 283	Ptolemy II founds the Library in Alexandria, with a famous collection of perhaps half a million books obtained by purchasing private libraries, including possibly Aristotle's. Astronomical instruments were constructed for use at the Library, and the matching of theory with observation was undertaken on a systematic and sustained basis.
ca. 200	Apollonius of Perga, born in the second half of the third century B.C., died early second century B.C. May have been a pupil of Euclid in Alexandria. Wrote the *Conics* on conic sections (parabola, hyperbola, and ellipse: curves cut from a right circular cone by a plane) and may well have devised the eccentric and epicycle hypotheses of planetary movement and demonstrated a proposition regarding the retrograde motions of the planets.
ca. 150	Hipparchus born in the first quarter of the second century B.C., died after 127 B.C. Developed a quantitative solar model and thus helped transform Greek geometrical astronomy from a qualitative to a quantitative science.

47	Apocryphal burning of the Library during Julius Caesar's occupation of Alexandria.
100 A.D.	Theon of Smyrna. Flourished early second century Author of a handbook containing citations from earlier sources illustrating interrelationships of arithmetic, geometry, music, and astronomy.
100–178	Ptolemy. He systematized and quantified with rigorous geometrical demonstrations and proofs hundreds of years of Greek geometrical astronomy, doing for astronomy what Euclid had done for geometry and earning for himself a reputation as the greatest astronomer of the ancient world.
140	Completion of Ptolemy's *Almagest,* as his mathematical systematic treatise of astronomy, *The Mathematical Syntaxis,* would come to be titled.
284	Roman Empire split into eastern and western halves.
330	Emperor Constantine transfers the capital of the Roman Empire to Byzantium.
392	In Alexandria the last fellow of the Museum is murdered by a mob and the Library is pillaged.
476	Fall of the western half of the Roman Empire.
500	Simplicius was born ca. 500 and died after 533 He wrote extensive commentaries on Aristotle's works, included within which was a statement of Plato's paradigm of uniform circular motion.
517	Commentary on Aristotle by John Philoponus.
529	Emperor Justinian banishes the Academy from Athens, sending scholars fleeing eastward to Persia, where Aristotelian studies would flourish, especially under Islamic civilization beginning a century later.
622	Mohammed arrives in Medina.
640	Muslim conquest of Alexandria.
670	Islam spreads through North Africa.
711	Islam spreads across the Mediterranean to Spain.
732	Muslim invasion of Europe halted at the Battle of Tours.
1126–1198	ibn-Rushd (Averroës). Outstanding Islamic astronomer, philosopher, and medical doctor among the Spanish Aristotelians.
1190	al-Bitruji (Alpetragius). Outstanding Islamic astronomer among the Spanish Aristotelians.

1200–1500	Scholasticism, a way of thinking, and a fusion of Aristotelianism and Christianity, permeates Western Europe, especially in universities.
1201–1274	Nasir al-Din al-Tusi. Famous Persian astronomer who devised the Tusi couple producing straight line motion from a combination of uniform circular motions, founded the Maragha Observatory, and probably influenced Copernicus.
1248	Christians retake Cordoba and Seville form the Muslims.
1258	Mongol invaders under Hulagu Khan, a grandson of Genghis Khan, conquer Baghdad.
1258	Founding of the Maragha Observatory in what is now northwestern Iran.
1277	Condemnation of 1277.
1295–1358	Jean Buridan. He taught natural philosophy at the University of Paris and wrote several commentaries on books by Aristotle. He proposed the idea of impetus.
1300–1500	Humanism. An intellectual movement overlapping Scholasticism and bringing a renewed interest in Plato. Neoplatonism, also called Neopythagoreanism, included a new belief in the possibility and importance of discovering simple arithmetic and geometric regularities in nature, and a new view of the Sun as the source of all vital principles and forces in the universe.
1300–1632	Renaissance; Renaissance humanism. Initially, a rebirth of Greek philosophy and values. Recovery, translation, and diffusion of lost classical works marked the first stage of the Renaissance and of humanism. The Renaissance began in Italy in the fourteenth century and spread to universities north of the Alps in the fifteenth and sixteenth centuries. Inconsistencies within individual ancient works and between different authors, and discrepancies in the sciences between classical theory and contemporary observation initially could be attributed to defects in transmission and translation. Eventually, however, critical thought was stimulated, and what had begun as a rebirth or recovery of old knowledge mutated into the creation of new knowledge.
1305–1375	Ibn al-Shatir. Distinguished Damascus astronomer, whose lunar theory when found in manuscripts

around the middle of the twentieth century was recognized as being nearly identical with Copernicus's.

1320–1382   Nicole Oresme. French natural philosopher. Discussed the hypothetical possibility of the rotation of the Earth.

1423–1461   Georg Peurbach. German astronomer and mathematician. Co-author with his pupil Regiomontanus of *Epitome Almagesti Ptolemaei*, the first Renaissance textbook on Ptolemy's astronomy.

1436–1476   Johannes Müller, known as Regiomontanus, after his home city. Pupil of Perubach; saw through to completion their new translation of Ptolemy's *Almagest.*

1453   Fall of Constantinople to the Turks and the end of the eastern half of the Roman Empire.

1473–1543   Nicholas Copernicus. Polish astronomer whose substitution of the Sun for the Earth in the center of the universe lead to a scientific revolution unforeseen by its author.

1492   Christians retake Granada from the Muslims. Christopher Columbus discovers a new world.

1496   Publication of Regiomontanus' *Epitome of the Almagest.*

1498–1552   Andreas Osiander. Lutheran clergyman who replaced Rheticus as editor of Copernicus's *De revolutionibus* and surreptitiously added an unsigned preface presenting Copernicus's theory as a mathematical fiction rather than the true account of the beauty and harmony of the universe that Copernicus intended.

1514   Copernicus distributes his *Commentariolus (Little Commentary)*, a sketch of his hypotheses of the heavenly motions. It soon disappeared from circulation and wasn't published until 1878 after a handwritten copy was found in Vienna.

1514–1574   George Rheticus. German mathematician, astronomer, and astrologist. Unable legally to use his own surname because his father had been beheaded for sorcery, he chose a name indicating his birthplace in what had been the ancient Roman province of Rhaetia. Wrote in 1540 the *Narratio prima*, or first report, on Copernicus's heliocentric system, and arranged for publication of Copernicus's *De revolutionibus.*

October 31 of 1517	Martin Luther posts his 95 theses as an academic challenge to a disputation and initiates the Reformation, a religious movement aimed at an internal renewal of the Catholic Church but resulting in a great revolt against it.
1539	Martin Luther criticizes the fool would turn the whole science of astronomy upside down, the new astronomer who wants to prove that the Earth goes round, and not the heavens, the Sun, and the Moon.
1540	Publication of Rheticus's *Narratio prima* (*First Account*), of part of Copernicus's theory.
1543	Publication of Copernicus's *De revolutionibus.*
1546–1601	Tycho Brahe. Observed the new star of 1572 and comet of 1577, changes in the supposedly unchanging Aristotelian heavens. Built more accurate astronomical instruments than any before, and his observations were used by Kepler to revolutionize astronomy.
1556	Leonard Digges, father of the English astronomer Thomas Digges, publishes a standard medieval diagram of the universe with the Earth in the center.
1564–1642	Galileo Galilei. Destroyed the Aristotelian worldview with his discoveries of mountains on the Moon, satellites of Jupiter, sunspots, and phases of Venus. He becomes entangled with religious questions concerning how to interpret the Bible and is now the stereotype for warfare between science and theology.
1571–1630	Johannes Kepler. Famous for his three laws, whose accounting for by Newton's theory of gravity would win over skeptics to Newton's worldview.
1572	New star of 1572.
1576	Tycho Brahe receives island of Hven from the King of Denmark, where he would build his observatory. The English astronomer Thomas Digges reprints his father Leonard Digges' book with a new appendix containing a diagram depicting stars scattered at varying distances beyond the former boundary of the sphere of the stars.
1577	Comet of 1577. Istanbul observatory torn down, its attempt to pry into the secrets of nature suspected of having brought on plague, defeats of Turkish armies, and the deaths of several important persons, all in the wake of the famous comet of 1577.

1582	Gregorian calendar, based on computations using the Copernican system, introduced by Pope Gregory VIII.
	Tycho Brahe completes his great mural quadrant at Uraniborg. It was forged of brass, 2 meters in radius, 13 millimeters broad, 5 millimeters thick, and marked in sixths of a minute.
1596–1650	René Descartes. French philosopher and scientist. In his *Discourse on the method of rightly conducting reason and seeking truth in the sciences,* he had begun with the famous phrase "Cogito ergo sum" (I think, therefore I am) and then moved, one step at a time, to include the existence of God, the reality of the physical world, and its mechanistic nature. His universe consisting of huge whirlpools, or vortices, of cosmic matter presented a serious rival to Newton's worldview.
1596	Publication of Kepler's *Mysterium cosmographicum.*
1597	Tycho Brahe leaves Hven.
1600	Tycho Brahe and Kepler united.
	William Gilbert, royal physician to both Elizabeth I and James I in England, publishes *De magnete,* his study of the magnet and magnetic bodies and that great magnet the Earth.
1609	Publication of Kepler's *Astronomia nova.*
	Rumor reaches Galileo of a device using pieces of curved glass to make distant objects on the Earth appear near.
1610	Publication of Galileo's *Starry Messenger.*
	Galileo takes up residence in Florence as mathematician and philosopher to the grand duke, and also as chief mathematician of the University of Pisa, without obligation to teach.
	Galileo observes phases of Venus.
1611	John Donne's poem *The Anatomy of the World* refers to Christian morality as well as the physical locations of the Sun and the Earth in noting that the new philosophy called all in doubt and that the heirarchy of prince, subject, father, son, were overthrown.
1612	Galileo begins a systematic study of sunspots.
1613	Publication of Galileo's *Letters on Sunspots.*
1614	John Napier, a Scottish nobleman, publishes *Mirifici logarithmorum canonis descriptio* (*Description of the Admirable Table of Logarithms*) one of the most useful arithmetic concepts in all science.

1615	Composition of Galileo's *Letter to Christina*.
1616	Galileo instructed neither to hold nor defend the heliocentric hypothesis.
1619	Publication of Kepler's *Harmonice mundi*.
1624	Galileo's friend and supporter Cardinal Maffeo Barberini became Pope Urban VIII.
1632	Galileo publishes his *Dialogue Concerning the Two Chief World Systems*. Galileo forced to abjure and sentenced to life imprisonment; immediately commuted to permanent house arrest under surveillance.
1637	Publication of René Descartes' *Le discours de la méthode pour bien conduire sa raison et chercher la vérité dans les sciences (Discourse on the method of rightly conducting reason and seeking truth in the sciences)*.
1638	Publication of *The Discovery of a New World: or, a Discourse tending to prove, that it is probable there may be another habitable World in the Moon*. John Wilkins, a major figure in the establishment of the Royal Society of London and its first secretary, guesses that there are some lunar inhabitants, even if there was no direct evidence. Why else would Providence have furnished the Moon with all the conveniences of habitation shared by the Earth?
1642	Death of Galileo and birth of Newton.
1642–1727	Isaac Newton. English mathematician, astronomer, physicist, and natural philosopher. Found the natural laws binding together the physical world and inspired others to search for natural laws in other realms, including but by no means limited to politics and economics.
1646–1716	Gottfried Leibniz. German mathematician. Disputed Newton for priority in the invention of the calculus and criticized theological implications of Newton's cosmology.
1656–1743	Edmond Halley. Predicted the return of his eponymous comet and was instrumental in the publication of Newton's *Principia*.
1657–1757	Bernard Fontenelle. French astronomer, mathematician, and writer, of a book, *Conversations on the Plurality of Worlds*, making science accessible to an educated but nonspecialized public and opening a market for women readers.

1666	While musing about apples falling, a flash of insight, so the legend goes, gave Newton the idea of universal gravitation.
1681	Samuel Colvil in *The Whigs Supplication* describes, as seen through a telescope, lunar inhabitants engaging in all the vices of Earth's society.
August 1684	Edmond Halley visits Newton and asks him what he thought the curve would be that would be described by the planets, supposing the force of attraction towards the Sun was the reciprocal to the square of their distances from the Sun.
November 1684	Newton sends Halley a nine page essay, *De motu corporum in gyrum* (*On the Motion of Bodies in an Orbit*).
1686	Bernard Fontenelle publishes his *Entretiens sur la pluralité des mondes* (*Conversations on the Plurality of Worlds*). The book was an instant best-seller, made it onto the Catholic index of prohibited books, and continues to be read today.
1687	Publication of one of the most influential and important books ever written, Newton's *Philosophiae naturalis principia mathematica* (*Mathematical Principles of Natural Philosophy*).
A.D 1695.	Edmond Halley realizes that the comets of 1531, 1607, and 1682 had similar orbits, and he predicts the return of this comet in 1758.
1698–1759	Pierre Maupertuis. French mathematician and physicist. He led an expedition to Lapland that measured flattening of the Earth toward the poles, as predicted by Isaac Newton, not flattening at the equator as predicted by Cartesians.
1701–1783	Leonhard Euler. Swiss mathematician. He helped develop necessary mathematical techniques to compute perturbation effects. First, he applied them to the Moon, and then, in 1748, to Jupiter and Saturn. The French Académie des Sciences' prize topic for 1748 called for an explanation of the inequalities of motion that Jupiter and Saturn appeared to cause in each other's motions, their observed motions having been found inexplicable solely on the basis of Kepler's laws and an inverse-square force of gravity. Euler extended his analysis of the interaction of Jupiter and Saturn in another prize essay, in 1752.
1718	Edmond Halley announces that a comparison of stellar positions in his day with those measured

by Hipparchus reveals that in the course of 1,800 years several of the supposedly fixed stars have altered their places relative to other stars.

1724–1804   Immanuel Kant. German philosopher. In his *Universal Natural History and Theory of the Heavens*, Kant attempted to extend Newtonian philosophy, as is indicated in the subtitle of the manuscript: *An Essay on the Constitution and Mechanical Origin of the Whole Universe Treated According to Newton's Principles*. Kant's manuscript perished in his printers' bankruptcy, and his cosmological ideas were not published until 1763, and then only in an appendix of another book.

1727   Death of Isaac Newton. Alexander Pope's famous epitaph reads: "Nature and Nature's Laws lay hid at Night; God said *Let Newton be!* and All was *Light*" (Feingold, *Newtonian Moment*, 144).

1730   The Swiss mathematician Johann Bernoulli wins a prize from the Académie des Sciences for his essay on the causes of elliptical orbits in a Cartesian vortex.

1733   Publication of Voltaire's *Lettres philosophiques*, also known as *Lettres anglaises*, or *Letters Concerning the English Nation*.

1737   Publication of Voltaire's *Elémens de la philosophie de Newton* (*Elements of Newton's Philosophy*).
Maupertuis returns to Paris with measurements supporting Newton and refuting Descartes.

1740   Maupertuis publicly declares his allegiance to Newtonian cosmology because of the failure of Cartesians to reconcile vortex theory with Kepler's mathematical laws of planetary motion.

1749–1827   Pierre-Simon Laplace. French mathematician, physicist, and celestial mechanic. Completed the Newtonian program of accounting for planetary motions as due to an inverse square force of gravity.

1750   The self-taught English astronomer Thomas Wright proposes as a model for the Milky Way a luminous band of light circling the heavens.

1755   Kant, inspired by an incorrect summary of Wright's book and by the paradigm of the Newtonian solar system, explains that the Milky Way is a disk-shaped system seen from the Earth located in the plane of the disk. Furthermore, nebulous patches of light in the Heavens are composed of stars and are other

	Milky Ways, or island universes (galaxies similar to and beyond the boundaries of our own galaxy).
1758	Return of Halley's comet, as he predicted before his death in 1743.
1761	Transit of Venus, as Halley predicted before his death in 1743.
1769	Transit of Venus, as Halley predicted before his death in 1743.
1776	Thomas Jefferson invokes laws of nature in the American Declaration of Independence.
	The Scottish economist Adam Smith invokes laws of nature in his *Wealth of Nations*.
1785	Laplace resolves a large anomaly in the motions of Jupiter and Saturn.
1787	Laplace resolves an anomaly in the acceleration of the Moon's orbital speed around the Earth.
1796	Publication of Laplace's *Exposition du système du monde*.
1799–1808	Publication of Laplace's *Mécanique céleste* in four volumes.
1810	Publication of *Zur Farbenlehre*, Johann Goethe's fierce attack on Newton's optics.
1861–1916	Pierre Duhem. French physicist and historian and philosopher of science. An ardent defender in partisan polemical battles of nineteenth-century French instrumentalism against rival views of scientific theory, Duhem found in medieval philosophy the birth of modern scientific theory. His instrumentalist view became the dominant characterization of astronomical science from the Greeks through Copernicus. Only recently have scholars came to favor a realist interpretation: the scientists thought they were discovering reality, not merely useful fictions. Yet another theme set by Duhem for the history of medieval science was his assertion that mechanics and physics proceeded by a series of small improvements from the Middle Ages to modern times. Some of Duhem's more extreme claims, including the purported anticipation of the concept of inertia in medieval impetus theory, have been modified. Duhem found in the condemnation of 1277 a blow against entrenched Aristotelianism and therefore an opening for the birth of modern science; others interpret the condemnations as a subordination of philosophy to theology.

1875	Speculative geometrical reconstruction of Eudoxus's system of homocentric spheres is published by the Italian astronomer Giovanni Schiaparelli, who would later become famous for his purported discovery of *canali* on Mars.
1887	Michelson-Morley experiment rules out any luminiferous light-bearing medium in the universe.
1892–1964	Alexandre Koyré. French historian of science, philosophy, and ideas. Examined philosophical issues in Isaac Newton's work and emphasized thought experiments rather than actual physical experiments in Galileo's work.
1893	Publication of a 1389 manuscript of al-Tusi's *Al-Tadhkira* found in the Bibliothèque Nationale in Paris.
1899–1990	Otto Neugebauer. German-American historian of ancient mathematical astronomy.
1900	Greek astronomical device from the first century B.C. is found by a sponge diver in an ancient shipwreck off the tiny Greek island of Antikythera.
1905–1983	Arthur Koestler. Hungarian journalist, novelist, political activist, social philosopher, and historian of science. In his 1959 book *The Sleepwalkers,* he described the history of cosmic theories as a history of collective obsessions and controlled schizophrenias, the scientists caught up in feverish intellectual dreams. Koestler's interest was in the psychological process of discovery and in the converse process that blinds a person towards truth which, once perceived, is heartbreakingly obvious. He highlighted unconscious biases and philosophical and political prejudices: the underlying beliefs, values, and worldviews that lie behind the quasi-aesthetic choices that scientists make. No branch of science, Koestler asserted, ancient or modern, can claim freedom from metaphysical bias of one kind or another. Biases against Koestler's then novel and radical view of science deterred potential followers.
1906	The astronomer and historian of astronomy J.L.E Dreyer publishes his *History of the Planetary Systems from Thales to Kepler.*
1908	The Romanian sculptor Constantin Brancusi, trying to make his art a working philosophy of Plato, abandons details and retains only essen-

tial elements in his sculptural kissing couple, who represent an ideal.

1922–1996   Thomas Kuhn. Historian of science. He questioned the common notion of scientific progress as a clean, rational advance along a straight, ascending line; and he shattered the reassuring vision of cumulative and progressive science with his 1962 book *The Structure of Scientific Revolutions*, one of the most exciting books ever written about science. Some radicals have seized on the book as a platform from which to oppose all cognitive authority. Scientists have objected vehemently to Kuhn's demonstration that personalities and politics play a role in science. Kuhn denies that there exists one full, objective, true account of nature, or that scientific achievement can be measured in terms of how close it brings us to truth.

1922–   Gerald Holton. Physicist and historian of science. For Holton, science is not so much a collection of facts, but rather is characterized by scientists thinking about and wrestling with problems. He tries to determine what observations and theories a scientist was confronted with in his own era, the state of shared scientific knowledge in that particular place and time, the creative insight that guided the scientist, and possible connections between the scientist's scientific work and his lifestyle, sociological setting, and cultural milieu.

1959   Publication of Arthur Koestler's *The Sleepwalkers*.

1962   Publication of Thomas Kuhn's *The Structure of Scientific Revolutions*.

1977   Robert Newton, a geophysicist at the Johns Hopkins University Applied Physics Laboratory, publishes *The Crime of Claudius Ptolemy*, in which he argues that Ptolemy was the most successful fraud in the history of science, and that the *Almagest* did more damage to astronomy than any other work ever written.

1978   Pope John Paul II acknowledges that Galileo's theology was sounder than that of the judges who condemned him.

1992   The pope sets up a special committee to reexamine the Galileo case, and offers an official apology for Galileo's sentence.

A. D. **1994**	Archaeologists recover pieces of the lighthouse form the harbor at Alexandria.
1996	An examination of hairs from Tycho Brahe's beard finds increased levels of mercury.
1999	Release of the movie *The Matrix*, a modern variation on Plato's cave.
2002	Completion of the new library at Alexandria. New analysis of the Greek astronomical device froun off Antikythera.
2004	Archaeologists find what they believe to be the site of the Library of Alexandria and a complex of 13 lecture halls.
2005	Discovery in the statue of Atlas at the National Archaeological Museum in Naples of a pictorial presentation of data from Hipparchus's long-lost star catalog. Archaeologists find what they believe to be Copernicus's grave and remains under the floor of Frombork Cathedral. Scientists will try to find relatives of Copernicus and do DNA identification.

GLOSSARY

A posteriori: a type of reasoning, from observation to theory, from facts or particulars to general principles, from effects to causes; synonymous with *inductive* and *empirical*. See *regular solids*.

A priori: a type of reasoning, from theory or principles without prior observations. See *regular solids*.

Anomaly: a violation of expectation; a discovery for which an investigator's paradigm has not prepared him or her.

Anthropocentric universe: centered on human interests. The Copernican revolution saw a historical progression from belief in a small universe with humankind at its center to a larger, and eventually infinite, universe with Earth not in the center. The physical geometry of our universe was transformed from geocentric and homocentric to heliocentric, and eventually to a-centric. The psychological change was no less. We no longer command unique status as residents of the center of the universe, enjoying our privileged place. Nor are we likely the only rational beings in the universe.

Antiperistasis: Aristotle's convoluted explanation for how an arrow shot from a bow, or a stone thrown from a hand, continued in motion with continuing contact between the moved and the mover. Somehow, air pushed forward by the arrow or the stone moved around to the rear of the arrow or stone and then pushed it from behind. See *impressed force*.

Aphelion: the point in an orbit farthest from the Sun (*apo*, "away from"; *helio*, "the Sun"). See *perihelion*.

Apogee: the point in an orbit farthest from the Earth (*gee*, "Earth"). See *perigee*.

Apple: a fruit growing on trees, and supposedly instrumental in several major historical changes, beginning with Adam, Eve, a serpent, and an apple; moving on to a golden apple awarded by the prince of Troy to the most beautiful woman in the world, who then promised him another man's wife, thus setting off the Trojan War; and Isaac Newton's apple, which, in contrast to the first two, rendered great service to humankind.

Astrotheology: the eighteenth-century belief in the orderliness of the universe combined with a belief that determination of that order was an important theological, philosophical, and scientific endeavor.

Chain: Adam Smith, initially believing that philosophical systems were mere inventions of the imagination to connect together otherwise disjointed and discordant phenomena of nature, was persuaded by the example of Isaac Newton's science that there are real connecting principles or chains that bind nature.

Circle: a closed curve everywhere the same distance from its center; acclaimed by ancient Greek mathematicians as the perfect geometrical figure. Supposedly, Plato set for astronomers the task of explaining the apparently irregular motions of the planets, the Sun, and the Moon as a combination of circular motions with constant speeds of rotation. See *saving the phenomena.*

Comet: a celestial body with a dense head and a vaporous tail. It is visible only if, and when, its orbit takes it near the Sun. Aristotle had made a distinction between the corrupt and changing sublunary world (from the Earth up to the Moon) and the perfect, immutable heavens beyond. Tycho Brahe's observations of the comet of 1577 showed that it was above the Moon and moving through regions of the solar system previously believed filled with crystalline spheres carrying around the planets. Aristotle's world was shattered. See *new star; crystalline sphere.*

Commentaries: ancient and medieval discussions or presentations based on books by Aristotle. Commentaries were neither modern nor scientific in spirit. Everyone knew what the questions were and what the answers would be. The aim of an exercise was skillful presentation of known information, not discovery of new information. The medieval university curriculum was based partly on studying commentaries on Aristotle's works on logic and natural philosophy.

Condemnation of 1277: Necessary principles can result in truths necessary to philosophy but contradictory to dogmas of the Christian faith. In 1270 the bishop of Paris condemned several propositions derived from the teachings of Aristotle, including the eternity of the world and the necessary control of terrestrial events by celestial bodies. In 1277 the Pope directed the bishop to investigate intellectual controversies at the university. Within three weeks, over two hundred propositions were condemned. Excommunication was the penalty for holding even one of the damned errors. Some historians assert that the scientific revolution of the sixteenth and seventeenth centuries owes much to the Condemnation of 1277. Though intended to contain and control scientific inquiry, the condemnation may have helped free cosmology from Aristotelian prejudices and modes of argument. But if so, why did scholars wait hundreds of years before repudiating Aristotelian cosmology?

Conic section: a curve produced from the intersection of a plane with a right circular cone: circle, ellipse, hyperbola, and parabola. Apollonius's book on conic sections, *Conics,* was recovered along with other ancient Greek classics in the early stages of the Renaissance. The circle and the ellipse are both conic sections, and for Kepler, neither geometrical figure may have seemed any more natural or more perfect than the other. And thus, maybe he could casually jettison circular orbits and some two thousand years of tradition.

Crisis: in science, the period of great uncertainty when an anomaly is judged worthy of concerted scrutiny yet continues to resist increased attention, with repeated failures to make the anomaly conform, and thus leads to large-scale paradigm destruction and major shifts in the problems and techniques of normal science. External social, economic, and intellectual conditions may also help transform an anomaly into a crisis.

Crystalline sphere: in ancient Greek and medieval European cosmology, solid crystalline spheres provided a physical structure for the universe and carried the planets in their motions around the Earth. Then, Copernicus moved Earth out of the center of the universe, making the spheres obsolete, and Tycho Brahe shattered them with his observations of the comet of 1577. See *comet*.

Cuneiform writing: wedge-shaped signs made by pressing a sharpened stylus or stick into soft clay tablets, often about the size of a hand. The tablets were then baked hard, preserving the contents, in some cases Babylonian astronomical knowledge.

Deduction: logical, often mathematical, derivation from theory of what phenomena may be expected. An inference of the sort that if the premises are true, the conclusion necessarily follows. See *hypothetico-deductive method; induction*.

Deferent: in ancient Greek geometrical astronomy, a large circle rotating at a constant speed around its center (coinciding with the Earth) and carrying around on its circumference the center of a smaller rotating circle, which in turn carried a planet on its circumference. See *eccentric; epicycle*.

Earthshine: sunlight reflected from the Earth. Near new moon, the portion of the Moon shadowed from direct sunlight is slightly brightened by sunlight reflected from the Earth.

Eccentric: a deferent circle with its center offset from the Earth. See *deferent*.

Eclipse: a partial or complete temporary blocking of light by an intervening body. A solar eclipse occurs when the Moon comes between the Earth and the Sun. A lunar eclipse occurs when the Earth is directly between the Moon and the Sun.

Ecliptic: the apparent path of the Sun among the stars; the intersection of the plane of the Earth's orbit around the Sun with the celestial sphere. See *zodiac*.

Ellipse: the closed curve generated by a point (the locus of the points) moving in such a way that the sum of the point's distances from two fixed points is a constant. Kepler showed (his first law) that the planets' orbits are not circles but ellipses. See *Kepler's first law*.

Enlightenment: the eighteenth-century philosophical movement concerned with critical and rational examination of previously accepted ideas and institutions.

Ephemeris: a tabular statement of the places of a celestial body at regular intervals. Seemingly, ancient Babylonian astronomers were content with tables of predicted celestial positions. We have no evidence that they constructed geometrical models of the motions of celestial bodies, or that they expressed concern about the causes of the motions or any curiosity about the physical composition of the celestial bodies, at least not in their clay tablets found and studied so far.

Epicycle: in ancient Greek geometrical astronomy, a small circle rotating at a constant speed around its center and carrying around a planet on its circumference; the center of the epicycle is carried around on the circumference of a larger rotating circle, the deferent.

Equant point: a geometrical invention of the ancient astronomer Ptolemy necessary to save the phenomena. Uniform angular motion, previously defined as cutting off equal angles in equal times at the center of the circle, was now taken with respect to this new point not at the center of the circle. The equant point was a questionable modification of uniform circular motion, and Copernicus would condemn the equant point as an unacceptable violation of uniform circular motion. See *saving the phenomena*.

Equator: the great circle around the Earth's surface defined by a plane passing through the Earth's center and perpendicular to its axis of rotation.

Equinox: either of the two points (the vernal, or spring, equinox, about March 21; and the autumnal, or fall, equinox, about September 22) on the celestial sphere where the plane of the Earth's orbit around the Sun and the plane of the Earth's equator intersect. At these times of the year the length of day and night are equal (12 hours) every place on the Earth.

Ex post facto argument: made after the fact, or observation, on which it is based. Also a theory modified to bring it into agreement with new facts. Such theories carry less psychological conviction than do those predicting previously unknown phenomena, even if the strict logic of the two situations is equally compelling.

Foundational: leading to. Kepler's observations did not lead to Newton's derivation of the inverse square law of gravity. Rather, establishment of the concept of universal gravitation enshrined Kepler's three laws among the great achievements of science. Kepler's laws were important, however, in the acceptance of Newton's theory.

Full moon: the phase of the Moon when its entire disk is seen illuminated with light from the Sun. This occurs when the Moon is opposite the Earth from the Sun. See *new moon*.

Gravity: Isaac Newton explained the planetary motions and tides on the Earth by the inverse square power of a mysterious entity called gravity, but he was unable to explain the cause of this power. He argued for setting aside the question of what gravity is and to be content with a mathematical description of its effects: "And to us it is enough that gravity does really exist, and act according to the laws which we have explained, and abundantly serves to account for all the motions of the celestial bodies, and of our sea" (*Principia* III, General Scholium).

Great chain of being: in philosophical thought, it linked God to man to lifeless matter in a world in which every being was related to every other in a continuously graded, hierarchical order. Governmental order reflected the order of the cosmos, and thus social mobility and political change were crimes against nature. This began to change with Copernicus.

Handbook tradition: Greek science was distilled into handbooks, and through this medium it became known to Latin readers. Ptolemy's *Almagest*, however, was written after the incorporation of Greek science into handbooks.

Harmony: an aesthetically pleasing combination of parts. Believing that God established nothing without geometrical beauty, Kepler compared the intervals between planets with harmonic ratios in music. See *symmetry*.

Humanism: a cultural and intellectual movement from roughly 1300 to 1600 centered on the recovery of ancient Greek knowledge. With humanism came a renewed interest in Plato. *Neoplatonism*, also called *Neopythagoreanism*, included belief in the possibility and importance of discovering simple arithmetical and geometrical regularities in nature, and a view of the Sun as the source of all vital principles and forces in the universe. Copernicus would rely on the recovery of Ptolemy's mathematical astronomy, for both its geometrical techniques and its philosophical human values. Eventually, critical thought was stimulated, and what had begun as a rebirth or recovery of old knowledge mutated into the creation of new knowledge.

Hypothetico-deductive method: First, a hypothesis is postulated; then, a prediction is deduced from the hypothesis; and finally, observations are made to determine if the phenomena deduced from the hypothesis exist. Note that hypotheses can be refuted but not proved.

Impetus: See *impressed force; inertia; momentum*.

Impressed force: an alternative to Aristotelian explanations of motion, which maintained contact between the mover and the moved (see *antiperistasis*). Some sort of incorporeal motive force could be imparted by a projector to a projectile. Planetary movements could be attributed to an impetus impressed by God at the creation. See *impetus; inertia; momentum*.

Induction: the reasoning process in which observations (phenomena) somehow are followed by the formation of theories (propositions). Science does not correspond to the inductive model in which all facts are collected and then inevitable theories are inevitably induced. This model fails because of an infinite number of possible observations. True inductive science would never advance beyond the infinite period of fact gathering to the stage of theory formulation. Modern science increasingly deviates from the inductive model. Theories increasingly are used to suggest observations of potential significance. In scientific discovery the catalytic role of intuition and hypothesis is essential in making sense of empirical results and in guiding further research. Note that unlike deduction, induction makes it possible for the premises to be true but the conclusion to be false. This is why it is possible logically to refute a theory but not to prove a theory. See *deduction*.

Inertia: the tendency of a body at rest to remain at rest, or if in motion to remain in motion. See *impetus; impressed force; momentum.*

Inferior planet: a planet lying between the Sun and the Earth (Mercury and Venus). The inferior planets can never be seen on the opposite side of the Earth from the Sun. See *opposition; superior planet.*

Instrumentalism: in the instrumentalist view of the relationship between theory and observation, known empirical data are suspended and the study then becomes one of pure mathematics, not solving but still relevant to the scientific problem. All that remains are simple mathematical fictions and pure conceptions, with no question of their being true or in conformity with the nature of things or even probable. For so-called instrumentalists, it is enough that a scientific theory yields predictions corresponding to observations. Theories are simply calculating devices. See *nominalism; realism.*

Kepler's first law: the planets move in elliptical orbits with the Sun at one focus; announced in Kepler's *Astronomia nova* (*New Astronomy*) of 1609, and discovered after his "second" law. That Tycho Brahe had set him to work on Mars's orbit Kepler later attributed to Divine Providence. Mars's orbit alone among the planetary orbits deviates so much from a circle that Tycho's observations would force Kepler from an eccentric circle to an oval. Then, Kepler found that the discrepancies between Tycho's observations and circular and oval orbits were equal and opposite, and that an elliptical orbit fit in between the circular and the oval. See *ellipse.*

Kepler's second law: or law of equal areas; also announced in Kepler's *Astronomia nova* (*New Astronomy*) of 1609, and discovered before his "first" law. The radius vector, the line from the Sun to the planet, sweeps out equal areas in equal times. If the areas of any two segments are equal, then the times for the planet to travel between the points on the orbit defining the two segments are also equal. Thus the distance of a planet from the Sun is inversely related to its orbital velocity: as the distance increases, the velocity decreases. That the planets move faster the nearer they are to the Sun had already been cited by Copernicus as a celestial harmony; Kepler now found further harmony in a quantitative formulation of the relationship. See *harmony.*

Kepler's third law: or harmonic law; announced in Kepler's *Harmonice mundi* (*Harmonies of the World*) in 1619. Among many propositions in this book on cosmic harmonies detailing various planetary ratios was the statement that the ratio of the mean movements of two planets is the inverse ratio of the $^3/_2$ powers of the spheres. Kepler's few readers could scarcely have guessed that this particular harmony would later be singled out for acclaim while all the other numerical relationships in the book would be discarded as nonsense, tossed into the garbage can of history. This third law is now usually stated as: the square of the period of time it takes a planet to complete an orbit of the Sun is proportional to the cube of its mean distance from the Sun. The law is also expressed as a ratio between two planets (A and B) going around the Sun (and also between two satellites going around a planet): the ratio of the periods squared is equal to the ratio of the distances cubed: (period A/period $B)^2 = $ (distance A/distance $B)^3$. See *harmony.*

Logarithm: The logarithm L to the base 10 of the number X is defined by the equation $X = 10^L$. Though it may not seem like a big deal to owners of electronic calculators, the introduction of logarithms early in the seventeenth century reduced the otherwise lengthy process of multiplication to simple addition and thus doubled Kepler's productivity and his working lifetime.

Momentum: a force of motion, the product of a body's mass times its linear velocity. See *impetus; impressed force; inertia.*

Neoplatonism: accompanied humanism, between 1300 and 1600, and included a renewed interest in Plato and a new belief in the possibility and importance of discovering simple arithmetical and geometrical regularities in nature, as well as a new view of the Sun as the source of all vital principles and forces in the universe. The movement was also called *Neopythagoreanism*, after the ancient Greek mathematician Pythagoras.

New moon: the phase of the Moon when none of its disk is seen illuminated with light from the Sun. This occurs when the Moon is between the Earth and the Sun. See *full moon.*

New star: a nova or supernova, a star that explodes and increases hundreds of millions of times in brightness. The famous nova of 1572 struck a blow against the Aristotelian worldview, in which there could be no change in the heavenly spheres beyond the Moon. If nearby, the nova would appear to shift its position with respect to the background stars. But Tycho Brahe showed that its parallax, or angle of view, did not change from night to night. See *comet.*

Nominalism: The nominalist thesis developed in the 1300s conceded the divine omnipotence of Christian doctrine but at the same time freed natural philosophy from religious authority. Science is a working hypothesis in agreement with observed phenomena. But we cannot insist on the truth of any particular working hypothesis. God could have made the world in some different manner but with the same set of observational consequences. Therefore scientific theories are tentative, not necessary, and can pose no challenge to religious authority. Nominalism has much in common with instrumentalism. Both philosophical concepts posit scientific theories as working hypotheses with no necessary links to reality. One might speculate that Catholic historians and philosophers of science in the twentieth century, justifiably impressed with fourteenth-century nominalism and also eager to praise the admirable achievements of that era by scholars working within and supported by the Catholic church, consequently were predisposed to formulate the concept of instrumentalism. Nominalism stripped of its religious context became instrumentalism, and nominalists were favorably pictured as forerunners of modern, philosophically sophisticated instrumentalists. See *instrumentalism.*

Normal science: the continuation of a research tradition, seeking facts shown by theory to be of interest.

Nova: see *new star.*

Opposition: a celestial object in opposition is located on the opposite side of the Earth from the Sun. The inferior planets, Mercury and Venus, can never be at opposition, because of the geometry of the situation. Hence, in the Ptolemaic Earth-centered model, the speeds around the Earth of the Sun and an inferior planet (Mercury or Venus) must be nearly matched to keep the planet and the Sun in approximately the same angular direction as seen from the Earth. In the Copernican model, an inferior planet is always at a small angle from the Sun. Nothing further is required of the model builder to obtain this result; it is a natural, inherent, and inevitable consequence of the model.

Paradigm: a universally recognized achievement temporarily providing model problems and solutions to a community of practitioners, telling scientists about the entities that nature contains and about the ways in which these entities behave.

Parallax: the difference in apparent direction of an object seen from different places. See *stellar parallax*.

Perfect solids: See *regular solids*.

Perigee: the point in an orbit nearest the Earth (*peri*, "near"; *gee*, "Earth"). See *apogee*.

Perihelion: the point in an orbit nearest the Sun (*helio*, "the Sun"). See *aphelion*.

Platonic solids: see *regular solids*.

Positivism: with regard to science, a philosophy including beliefs that science is cumulative and consists of a logical structure of testable and thus objective statements about a real world independent of investigators' personal subjective beliefs and general culture.

Precursitis: an imaginary disease involving the unconscious assumption that ancient scientists were working on the same problems and using the same methods as modern scientists do today. Hence the search for ancient precursors of the observations and theories now acclaimed in textbooks and the myopic result: a chronology of cumulative, systematized positive knowledge.

Pythagorean solids: See *regular solids*.

Quadrature: when the Moon is at 90 degrees from the line joining the Sun and the Earth.

Qualitative: not precise. See *quantitative*.

Quantitative: involving measurement and numbers. Throughout the eighteenth century, Newtonians piled quantitative success on quantitative success, mathematically explaining phenomena as the result of an inverse square force, while rival Cartesians failed to reconcile their vortex theory in any numerical detail with Kepler's mathematical laws of planetary motion.

Realism: an insistence that scientific theories are descriptions of reality. Dogmatic realists insist on the truth of a theory. Critical realists concede a theory's conjectural character without necessarily becoming instrumentalists. A disappointed realist may appear to be a local instrumentalist with regard to a particular failed theory retaining instrumental value but is far from becoming a global instrumentalist. See *instrumentalism*.

Regular solids: geometrical solids, each with all its sides equal, all its angles equal, and all its faces identical. There are five regular (or perfect or Platonic or Pythagorean) solids; no more, no less. They are the tetrahedron (4 triangular sides), cube (6 square sides), octahedron (8 triangular sides), dodecahedron (12 pentagonal sides), and icosahedron (20 triangular sides). About the Earth's circle, Kepler circumscribed a dodecahedron; enclosed the dodecahedron with Mars's circle; circumscribed about Mars's circle a tetrahedron; enclosed the tetrahedron with Jupiter's circle; about Jupiter's circle circumscribed a cube; and enclosed the cube with Saturn's circle. Within the Earth's circle, Kepler inscribed an icosahedron; in the icosahedron, he inscribed Venus' circle; in this circle, he inscribed an octahedron; and inside the octahedron, he inscribed Mercury's circle. The coincidence could not be purely fortuitous, could it? He used six planets (all that were known then), five intervals between them, and five regular solids! Kepler proclaimed this discovery in his *Mysterium cosmographicum* (*Cosmic Mystery*) in 1596. Kepler's theory was perceived as a mystical a priori speculation. It had originated more in his imagination than in observation. His order of work was preposterous. The proper procedure decreed for astronomers (however much they believed a priori in uniform circular motion) was to derive the distances of the planets a posteriori from observations, not a priori from the geometry of regular solids.

Renaissance: the rebirth (re-nascence) of Greek classical culture that originated in Italy in the fourteenth century and spread to universities north of the Alps in the fifteenth and sixteenth centuries. The University of Cracow was one of the first northern European schools to teach Renaissance humanism, although astronomy was still taught there largely in terms of Aristotelian physics in Copernicus's time. At the University of Bologna, Copernicus studied with Domenico da Novara, an astronomy professor and one of the leaders in the revival of Platonic and Pythagorean thought and Greek geometrical astronomy. Ptolemy's *Almagest* had become available to scholars in the Latin-reading world late in the fifteenth century. Initially, Renaissance humanists looking to the past for knowledge from a higher civilization facilitated a rebirth of Greek philosophy and values. Inconsistencies within individual ancient works and between different authors, and discrepancies in the sciences between classical theory and contemporary observation were attributed to defects in transmission and translation. Eventually, however, what had begun as a rebirth or recovery of old knowledge mutated into the creation of new knowledge. Obviously, any ending date for the Renaissance is arbitrary; some choose 1632 and the trial of Galileo.

Revolution-making: as opposed to revolutionary. Thomas Kuhn argued that Copernicus's work was almost entirely within an ancient astronomical tradition and hence not revolutionary, but it contained a few novelties that would lead to a scientific revolution and hence was revolutionary. See *revolutionary; scientific revolution*.

Revolutionary: See *revolution-making; scientific revolution*.

Saving the appearances: See *saving the phenomena*.

Saving the phenomena: in the context of ancient Greek geometrical astronomy, to devise a system of uniform circular motions that reproduced the observed phenomena, or appearances. The guiding themata or paradigm of

Greek planetary astronomy was attributed to Plato by the philosopher Simplicius of Athens in his commentary on Aristotle's book *On the Heavens*. Around 500 Simplicius wrote that Plato had set as a task for astronomers to explain the apparently irregular motions of the planets, the Sun, and the Moon as a combination of circular motions with constant speeds of rotation.

Scholasticism: a fusion of Aristotelianism and Christian theology permeated thought in Western Europe between roughly 1200 and 1500, especially in universities.

Scientific revolution: an extraordinary episode during which scientific beliefs, values, and worldviews are abandoned, and ruling paradigms are replaced by incompatible or incommensurable new paradigms. The Ptolemaic and Copernican systems are an example of incompatible systems. While they predicted different results, such as the appearance of the phases of Venus, they were judged by the mutual standard of how well each saved the phenomena with systems of uniform circular motions. Descartes' vortex system and Newton's gravity were more incommensurable than incompatible. They had different goals and different measurements of success or failure. Descartes insisted on an explanation of the cause of gravity, while Newton abandoned that quest and argued instead that it was enough that gravity acted according to an inverse square force law and accounted for all the motions of the celestial bodies. Though Descartes' later followers eventually agreed that any successful vortex theory would have to account for Kepler's laws, Descartes himself had ignored Kepler's discoveries, if indeed he had known of them. Newton, on the other hand, asked that his theory be judged by its success in accounting for Kepler's laws. Descartes' and Newton's theories were not incompatible, with a mutual standard against which they could be judged, but incommensurable: not comparable on any mutually agreed upon basis. Some historians accept the replacement of one theory with a second, incompatible theory as a revolution. Others might withhold the word *revolution* for the replacement of one worldview with another worldview so incommensurable that rival proponents cannot agree on common procedures, goals, and measurements of success or failure.

Sexagesimal number system: based on the number 60. Our time system of 60 minutes in an hour and 60 seconds in a minute is an example. Ancient astronomical positions were reported as 28, 55, 57, 58, with each succeeding unit representing so many sixtieths of the preceding unit (i.e., 28 degrees, 55 minutes, 57 seconds, etc.).

Sidereal month: the period of revolution of the Moon with respect to the stars. See *synodic month*.

Solstice: the times of the year when the Sun is at its highest or lowest latitude as seen from the Earth. Summer solstice is about June 22, and winter solstice is about December 22.

Stellar parallax: the angle subtended by the radius of the Earth's orbit at its distance from a star; the angle that would be subtended by one astronomical unit (the mean distance of the Earth from the Sun) at the distance of the star from the Sun. Stellar parallax was predicted from Copernicus's heliocentric theory but was too small to detect because the stars are at great distances from the Earth. Not until the 1830s, with instruments vastly better than the best in Copernicus's time, was stellar parallax first measured. See *parallax.*

Step function: Some Babylonian astronomers employed mathematical tables in which the solar velocity was represented as constant for several months, after which the Sun proceeded with a different constant speed for several more months before reverting to the initial velocity and remaining at that speed for several more months. Graphed, the motion would look like a series of steps up and down, and it is called a "step" function, although the Babylonians are not known to have used graphs. See also *zigzag function.*

Sunspots: dark spots appearing on the face of the Sun, associated with magnetic fields, and having a 22-year cycle from minimum to maximum activity and back to minimum activity.

Superior planet: a planet lying beyond the Earth outward from the Sun (Mars, Jupiter, etc.). See *inferior planet.*

Syllogism: a form of deductive reasoning. Aristotle established the use of syllogisms in logical presentations. He explained that a deduction is a discourse in which, certain things being stated, something other than what is stated follows of necessity. An example of a syllogism is the following: All organisms are mortal; All men are organisms; Therefore all men are mortal. Demonstrative syllogisms derive facts already known, not new facts. See *deduction.*

Symmetry: a relationship between constituent parts, often involving a sense of beauty as a result of harmonious arrangement. Plato regarded heaven itself and the bodies it contained as framed by the heavenly architect with the utmost beauty of which such works were susceptible. Ptolemy contemplated many beautiful mathematical theories, which lifted him from the Earth and placed him side by side with Zeus. Copernicus found an admirable symmetry in the universe and a clear bond of harmony in the motion and magnitude of the spheres. Could there possibly be a genetic wiring of the human brain, shaping our requirements for an aesthetically satisfying understanding of nature? See *harmony.*

Synodic month: the period of revolution of the Moon with respect to the Sun. See *sidereal month.*

Syzygy: when the Moon lies in a straight line with the Earth and the Sun, at opposition or conjunction.

Themata: underlying beliefs, values, and worldviews constraining or motivating scientists and guiding or polarizing scientific communities.

Tusi couple: an innovative combination of uniform circular motions producing motion in a straight line; devised by the Islamic astronomer Nasir al-din al-Tusi in the thirteenth century. See *saving the phenomena.*

Uniform circular motion: moving in a circle at a constant angular motion with respect to the center of the circle; Plato purportedly laid down the principle that the heavenly body's motions were circular, uniform, and constantly regular.

Vortices: huge whirlpools of cosmic matter. According to René Descartes, our solar system was one of many vortices, its planets all moving in the same direction in the same plane around a luminous central body. Planets' moons were swept along by the planets' vortices. All change in motion was the result of percussion of bodies; gravity was the result of celestial matter circulating about the Earth and pushing all terrestrial matter toward the Earth.

Zigzag function: a decreasing, increasing, and again decreasing sequence of numbers in a Babylonian table of astronomical positions, so called after how it would appear on a graph. Note, however, that the Babylonians are not known to have used graphs. See *step function*.

Zodiac: a band around the celestial sphere in which the Sun, the Moon, and the planets appear to move. It includes the intersection of the plane of the Earth's orbit around the Sun with the celestial sphere. See *ecliptic*.

ANNOTATED BIBLIOGRAPHY

PRIMARY SOURCES

Aristotle, *De caelo*, J. L. Stocks, trans., in W. D. Ross, ed., *The Works of Aristotle, vol. 2* (Oxford: Clarendon Press, 1930), reprinted, with mild emendations, in Jonathan Barnes, ed., *The Complete Works of Aristotle, vol. I* (Princeton: Princeton University Press, 1984). In *On the Heavens* Aristotle develops ideas raised in his *Physics*.

Aristotle, *Metaphysics; newly translated as a postscript to natural science, with an analytical index of technical terms, by Richard Hope* (New York: Columbia University Press, 1952).

Aristotle, *Physica*, in *Aristotle's Physics newly translated by Richard Hope with an Analytical Index of Technical Terms* (Lincoln: University of Nebraska Press, 1961). Aristotle's physics dominated scientific thought in the Western world for nearly two thousand years.

Copernicus, Nicholas, *Commentariolus,* in Edward Rosen, ed., *Three Copernican Treatises: The Commentariolus of Copernicus, The Letter against Werner, and the Narratio Prima of Rheticus* (New York: Dover Publications, 1939); *Second Edition, Revised, with an Annotated Copernicus Bibliography 1939–1958* (New York: Dover Publications, 1959). See also Noel Swerdlow, "The Derivation and First Draft of Copernicus' Planetary Theory: A Translation of the *Commentariolus* with Commentary," *Proceedings of the American Philosophical Society, 117* (1973), 423–512. Copernicus's early astronomical thoughts, in English translation and with extensive notes.

————, *De revolutionibus orbium caelestium.* Copernicus's great work of 1543 on the Revolutions of the Celestial Spheres. The preface and book I are conveniently available in English translation in Michael J. Crowe, *Theories of the World from Antiquity to the Copernican Revolution* (New York: Dover

Publications, Inc., 1990). Crowe selected his excerpts from *On the Revolutions: Nicholas Copernicus; translation and commentary by Edward Rosen* (Baltimore: Johns Hopkins University Press, 1978). Much of the preface and book 1 are also presented and commented extensively upon in Thomas Kuhn, *The Copernican Revolution: Planetary Astronomy in the Development of Western Thought* (Cambridge, Massachusetts: Harvard University Press, 1957). Kuhn took his passages from *De Revolutionibus, preface and book I,* John F. Dobson, trans., assisted by Selig Brodetsky (London: Royal Astronomical Society, 1947), originally printed as *Occasional Notes, no. 10,* May 1947. Other English translations of Copernicus's *De revolutionibus orbium caelestium* include *On the Revolution of the Heavenly Spheres,* Charles Glenn Wallis, trans., in *Great Books of the Western World, vol. 16: Ptolemy, Copernicus, Kepler* (Chicago: Encyclopedia Britannica, Inc., 1939; Buffalo, New York: Prometheus Books, 1993); and *On the Revolutions of the Heavenly Spheres; A New Translation from the Latin, with an Introduction and Notes by A.M. Duncan* (Newton Abbot: David & Charles, 1976; New York: Barnes & Noble, 1976). See also *As the World Turned: A Reader on the Progress of the Heliocentric Argument from Copernicus to Galileo.* [http://math.dartmouth.edu/~matc/Readers/renaissance.astro/0.intro.html.] Contains excerpts from Nicholas Copernicus, *De Revolutionibus;* John Dee, *The Mathematicall Praeface;* Robert Recorde, *The Castle of Knowledge;* Marcellus Palingenius Stellatus, *The Zodiake of Life;* Giordano Bruno, *The Ash Wednesday Supper; and* Galileo Galilei, *Dialogue Concerning the Two Chief World Systems.*

Fontenelle, Bernard le Bovier de, *Entretiens sur la pluralité des mondes,* in English translation with a useful introduction, notes, and bibliography in Bernard le Bovier de Fontenelle, *Conversations on the Plurality of Worlds, translation by H.A. Hargreaves, Introduction by Nina Rattner Gelbert* (Berkeley: University of California Press, 1990). Fontenelle imagined evening promenades in a garden with a lovely young marquise discussing that the Moon was inhabited, that the planets were also inhabited, and that the fixed stars were other suns, each giving light to their own worlds.

Galilei, Galileo, *Dialogue Concerning the Two Chief World Systems—Ptolemaic & Copernican. Translated, with revised notes, by Stillman Drake, foreword by Albert Einstein, second edition* (Berkeley: University of California Press, 1967). See also Maurice A. Finocchiaro, *Galileo on the World Systems* (Berkeley: University of California Press, 1997). The *Dialogue* plus texts related to Galileo's famous trial are available online in English translation at http://www.law.umkc.edu/faculty/projects/ftrials/galileo/galileo.html. Excerpts from the *Dialogue* are also available in *As the World Turned: A Reader on the Progress of the Heliocentric Argument from Copernicus to Galileo* [http://math.dartmouth.edu/~matc/Readers/renaissance.astro/0.intro.html]. It contains excerpts from Nicholas Copernicus, *De Revolutionibus;* John Dee, *The Mathematicall Praeface;* Robert Recorde, *The Castle of Knowledge;* Marcellus Palingenius Stellatus, *The Zodiake of Life;*

Giordano Bruno, *The Ash Wednesday Supper; and* Galileo Galilei, *Dialogue Concerning the Two Chief World Systems.*

————, *Sidereus nuncius.* Galileo's telescopic discoveries in his *Starry Messenger* of 1610, his 1613 letters on sunspots, and his *Letter to the Grand Duchess Christina* of 1615 discussing the relationship between science and religion are available in English translation in Stillman Drake, *Discoveries and Opinions of Galileo, translated with an introduction and notes by Stillman Drake* (Garden City, New York: Doubleday & Co., 1957). See also Albert van Helden, *Sidereus Nuncius or The Sidereal Messenger Galileo Galilei. Translated with introduction, conclusion, and notes by Albert van Helden* (Chicago: University of Chicago Press, 1989), which contains a good bibliography and discussion of Galileo's discovery of the phases of Venus.

Kepler, Johannes, *Astronomia nova.* The *New Astronomy* of 1609 is available in English translation in William H. Donahue, translator, *New Astronomy / Johannes Kepler* (Cambridge: Cambridge University Press, 1992).

————, *Epitome astronomiae Copernicanae.* Books IV and V of the *Epitome of Copernican Astronomy* of 1617–1621 are available in English translation in Charles Glenn Wallis, trans., *The Harmonies of the World: V,* in *Great Books of the Western World, vol. 16: Ptolemy, Copernicus, Kepler* (Chicago: Encyclopedia Britannica, Inc., 1939); reprinted in *Epitome of Copernican Astronomy & Harmonies of the World* (Amherst, New York: Prometheus Books, 1991).

————, *Harmonices mundi.* The *Harmonies of the World* of 1619 is available in English translation in *The Harmony of the World / by Johannes Kepler; translated into English with an introduction and notes by E. J. Aiton, A. M. Duncan, and J. V. Field* (Philadelphia: American Philosophical Society, 1997; *Memoirs of the American Philosophical Society, 209*). Book V is also available in Charles Glenn Wallis, trans., *The Harmonies of the World: V,* in *Great Books of the Western World, vol. 16: Ptolemy, Copernicus, Kepler* (Chicago: Encyclopedia Britannica, Inc., 1939); reprinted in *Epitome of Copernican Astronomy & Harmonies of the World* (Amherst, New York: Prometheus Books, 1991).

————, *Mysterium cosmographicum.* The *Cosmic Mystery* of 1596 is available in English translation in *Mysterium Cosmographicum = The Secret of the Universe* (New York: Abaris Books, 1981) by A.M. Duncan, trans., introduction and commentary by E.J. Aiton; with a preface by I. Bernard Cohen.

Newton, Isaac, *Philosophiae naturalis principia mathematica.* Among many English translations of the *Mathematical Principles of Natural Philosophy* of 1686 are the following: *Mathematical principles of natural philosophy and his system of the world by Sir Isaac Newton; translated into English by Andrew Motte in 1729. The translation revised and supplied with an historical and explanatory appendix, by Florian Cajori* (Berkeley: University of California Press, 1946); *Great Books of the Western World, vol. 34* (Chicago: Encyclopedia Britannica, Inc., 1955); S. Chandrasekhar, *Newton's Principia for the Common Reader* (New York: Oxford University Press, 1995); *Newton's Principia: the Central Argument: Translation, notes, and expanded proofs by Dana Densmore; Translations and diagrams by William H. Donahue* (Santa Fe, New Mexico: Green Lion Press, 1996); *The Principia: Mathematical Principles of Natural Philosophy. By Isaac Newton; A new translation by I. Bernard Cohen and Anne Whitman assisted by Julia Budenz; Preceded by a Guide to Newton's Principia by I. Bernard Cohen* (Berkeley: University of California Press, 1999). All of Newton's books published in his lifetime in Latin, English, and French are available online at: http://dibinst.mit.edu/BURNDY/Collections/Babson/Online-Newton/OnlineNewton.htm.

Plato, *Epistles*. In Glenn Morrow, *Plato's Epistles: A Translation, with Critical Essays and Notes, revised edition* (Indianapolis: Bobbs-Merrill, 1962). Contains Plato's letter "VII: To the Friends and Followers of Dion" with some of the difficulties Plato encountered in his eventful life, and also his definition of a circle. The letter's authenticity is not certain: see Morrow, "The Question of Authenticity," *ibid.*, pp. 3–6; and Paul Friedlander, *Plato: I An Introduction, second edition, with revisions,* translated by Hans Meyerhoff (Princeton: Princeton University Press, 1969; *Bollingen Series LIX*), pp. 236–245.

————, *Republic*. In Francis MacDonald Cornford, *The Republic of Plato* (New York: Oxford University Press, 1945). Contains Plato's Allegory of the Cave, which examines the nature of reality, and Plato's comments on the study of astronomy, which follow from his conception of reality. See Gregory Vlastos, "Degrees of Reality in Plato," in R. Bambrough, ed., *New Essays on Plato and Aristotle* (London: Routledge & Kegan Paul, 1965), 1–19; Vlastos, "The Role of Observation in Plato's Conception of Astronomy," in J. Anton, ed., *Science and the Sciences in Plato* (New York: Caravan Books, 1980), pp. 1–30; Norriss S. Hetherington, "Plato and Eudoxus: Instrumentalists, Realists, or Prisoners of Themata?," *Studies in History and Philosophy of Science, 27* (1996), 271–289; Hetherington, "Plato's Place in the History of Greek Astronomy: Restoring *both* History *and* Science to the History of Science," *Journal of Astronomical History and Heritage, 2* (1999), 87–110. This study examines Plato's philosophical vision and its influence on ancient science, and also casts new light on the long-standing debate over whether ancient Greek astronomers were instrumentalists or realists

Ptolemy, *Almagest*. In *Great Books of the Western World, vol. 16* (Chicago: Encyclopedia Britannica, 1952), R. Catesby Taliaferro, trans. Also available in English translation in G. J. Toomer, *Ptolemy's Almagest* (London: Duckworth, 1984; New York: Springer-Verlag, 1984). See also Toomer's encyclopedic entry "Hipparchus," in C. C. Gillispie, ed., *Dictionary of Scientific Biography, vol. 15* (New York: Charles Scribner's Sons, 1978), pp. 207–234. Scholars have generally accepted Toomer's harsh criticism of Taliaferro's translation and replaced it with Toomer's translation in their analyses, with gains perhaps in clarity and consistency but not in substantive conclusions. On internal inconsistencies in Ptolemy's system and why the Copernican revolution was so long delayed, see Norriss S. Hetherington and Colin Ronan, "Ptolemy's Almagest: Fourteen Centuries of Neglect," *Journal for the British Astronomical Association, 94:6* (October 1984), 256–262.

Rheticus, *Narratio Prima.* The first published account, by a young German professor of mathematics, Rheticus, of part of Copernicus's *De revolutionibus.* In English translation and with extensive notes, in Edward Rosen, ed., *Three Copernican Treatises: The Commentariolus of Copernicus, The Letter against Werner, and the Narratio Prima of Rheticus* (New York: Dover Publications, 1939); *Second Edition, Revised, with an Annotated Copernicus Bibliography 1939–1958* (New York: Dover Publications, 1959).

Simplicius of Athens, *Commentary on Aristotle's On the Heavens.* Simplicius's description of the problem that Plato set for astronomers is available in Pierre Duhem, *To Save the Phenomena: An Essay on the Idea of Physical Theory from Plato to Galileo,* translated by E. Doland and C. Maschler (Chicago: University of Chicago Press, 1969). Simplicius's description of the failure of Eudoxus's theory to save the phenomena is available in Thomas Heath, *Aristarchus of Samos: The Ancient Copernicus. A History of Greek Astronomy to Aristarchus together with Aristarchus's Treatise on the Sizes and Distances of the Sun and Moon. A New Greek Text with Translation and Notes* (Oxford: Clarendon Press, 1913; New York: Dover Publications, 1981), pp. 221–223. Some scholars doubt Simplicius's authority and question Plato's putative role in the development of planetary theory: Bernard R. Goldstein and Alan C. Bowen, "A New View of Early Greek Astronomy," *Isis, 74* (1983), 330–340; reprinted in Goldstein, *Theory and Observation in Ancient and Medieval Astronomy* (London: Variorum Reprints, 1985), pp. 1–11. Regarding the extent to which Eudoxus was beholden to Plato for the essential principle of his astronomical scheme, perhaps Plato envisioned oscillating motions for the planets and it was only later commentators who saw adumbrations of later astronomical models in Plato's earlier cosmological accounts: Wilbur R. Knorr, "Plato and Eudoxus on the Planetary Motions," *Journal for the History of Astronomy, 21* (1990), 313–329; and "Plato's Cosmology," in Norriss S. Hetherington, ed., *Encyclopedia of Cosmology: Historical, Philosophical, and Scientific Foundations of Modern Cosmology* (New York: Garland Publishing, 1993), pp. 499–502.

Smith, Adam, *An Inquiry into the Nature and Causes of the Wealth of Nations* (Edinburgh: 1776), available in many printed editions and online at http://www.adamsmith.org/smith/won-intro.htm. As Newton's *Principia* changed understanding of the physical world and Darwin's *Origin of Species* changed understanding of the biological world, so Smith's *Wealth of Nations* changed understanding of the economic world. For how Smith's work was inspired by Newton's, see Norriss S. Hetherington, "Isaac Newton and Adam Smith: Intellectual Links Between Natural Science and Economics," in Paul Theerman and Adele F. Seeff, *Action and Reaction: Proceedings of a Symposium to Commemorate the Tercentenary of Newton's* Principia (Newark: University of Delaware Press, 1993), pp. 277–291.

Theon of Smyrna, *Expositio rerum mathematicarum ad legendum Platonem utilium.* A handbook of citations from earlier sources on arithmetic, music, and astronomy intended to provide an exposition of mathematical subjects useful for the study of Plato. Excerpts with regard to the equivalence of the eccentric and epicyclic hypotheses are available in Pierre Duhem, *To Save the Phenomena: An Essay on the Idea of Physical Theory from Plato to Galileo,* translated by E. Doland and C. Maschler (Chicago: University of Chicago Press, 1969), pp. 8–11.

SECONDARY SOURCES

Aiton, E. J., *The Vortex Theory of Planetary Motions* (London: Macdonald & Co., 1972). On Descartes' vortex theory, the primary rival of Newton's theory of gravitation. See also B. S. Biagre, "Descartes's Mechanical Cosmology," in Norriss S. Hetherington, ed., *Encyclopedia of Cosmology: Historical, Philosophical, and Scientific Foundations of Modern Cosmology* (New York: Garland Publishing, Inc., 1993), pp. 164–176.

Andrade, E. N. da C., *Sir Isaac Newton: His Life and Work* (Garden City, New York: Doubleday & Co., Inc., 1954). Brief, popular historical introduction.

Armitage, Angus, *Edmond Halley* (London: Thomas Nelson and Sons Ltd., 1966). A scientific biography emphasizing Halley's scientific accomplishments.

———, *The World of Copernicus* (New York: Henry Schuman, Inc., 1947). Original title: *Sun, Stand Thou Still.* Brief, popular historical introduction.

Brush, Stephen, "Should the History of Science Be X-Rated?" *Science, 183* (1974): 1164–72.

Buchdahl, Gerd, *The Image of Newton and Locke in the Age of Reason* (London: Sheed and Ward, 1961). Small booklet on Newton's image and its effect on eighteenth-century thought and imagination.

Casper, Max, *Kepler, translated and edited by C. Doris Hellman* (London and New York: Abelard-Schuman, 1959). The most extensive biography by the greatest Kepler scholar.

Cohen, I. Bernard, *The Birth of a New Physics, Revised and Updated* (New York: W. W. Norton & Company, 1985). Expanded version of the 1960 original. An outstanding book showing clearly how Galileo, Kepler, and Newton changed Aristotle's physics.

Cook, Alan, *Edmond Halley: Charting the Heavens and the Seas* (Oxford: Clarendon Press, 1998). More detailed than Armitage's earlier biography.

Crowe, Michael J., *Theories of the World from Antiquity to the Copernican Revolution* (New York: Dover Publications, Inc., 1990). Outstandingly clear presentation, developed from a college course.

Dicks, D.R., *Early Greek Astronomy to Aristotle* (Ithaca, New York: Cornell University Press, 1970). Chapters on Eudoxus and on Callippus and Aristotle.

Drake, Stillman, *Dialogue Concerning the Two Chief World Systems—Ptolemaic & Copernican. Translated, with revised notes, by Stillman Drake, foreword by Albert Einstein, second edition* (Berkeley: University of California Press, 1967). See also Finocchiaro's translation.

———, *Discoveries and Opinions of Galileo, translated with an introduction and notes by Stillman Drake* (Garden City, New York: Doubleday & Co., 1957). Galileo's telescopic discoveries in his *Starry Messenger* of 1610, his 1613 letters on sunspots, and his *Letter to the Grand Duchess Christina* of 1615 discussing the relationship between science and religion in English translation with historical introduction. Drake is a major Galileo scholar, and all of his many books on Galileo are good reading.

Dreyer, J.L.E., *History of the Planetary Systems from Thales to Kepler* (Cambridge: Cambridge University Press, 1906); *second edition*, revised by William Stahl, with a supplementary bibliography, and retitled *A History of Astronomy from Thales to Kepler* (New York: Dover Publications, 1953). Still useful reading after a century.

Duhem, Pierre, *To Save the Phenomena: An Essay on the Idea of Physical Theory from Plato to Galileo*, translated by E. Donald and C. Machler (Chicago: University of Chicago Press, 1969).

Evans, James, *The History and Practice of Ancient Astronomy* (New York: Oxford University Press, 1998). Focuses on the concrete details of ancient astronomical practice. See also by Evans "Ptolemaic Planetary Theory," in Norriss S. Hetherington, ed., *Encyclopedia of Cosmology: Historical, Philosophical, and Scientific Foundations of Modern Cosmology* (New York: Garland Publishing, 1993), pp. 513–526; "Ptolemy," in Hetherington, ed., *Cosmology: Historical, Literary, Philosophical, Religious, and Scientific Perspectives* (New York: Garland Publishing, 1993), pp. 105–145.

Feingold, Mordechai, *The Newtonian Moment: Isaac Newton and the Making of Modern Culture* (New York and Oxford: The New York Public Library/ Oxford University Press, 2004). The best single volume on Newton's effect on modern culture and thought. This is the companion volume to a library exhibition; see http://www.nypl.org/research/newton/index.html.

Finocchiaro, Maurice A., *Galileo on the World Systems. A New Abridged Translation and Guide* (Berkeley: University of California Press, 1997). See also Drake's translation.

———, *The Galileo Affair: A Documentary History* (Berkeley: University of California Press, 1989). The story is continued in *Retrying Galileo 1633–1992* (Berkeley: University of California Press, 2005).

Gillispie, C.C., *Pierre-Simon Laplace 1749–1827: A Life in Exact Science* (Princeton: Princeton University Press, 1997). On Laplace, who resolved many outstanding problems in the Newtonian paradigm. See also Roger Hahn, *Pierre Simon Laplace 1749–1827: A Determined Scientist* (Cambridge, Massachusetts: Harvard University Press, 2005).

Gingerich, Owen, *The Eye of Heaven: Ptolemy, Copernicus, Kepler* (New York: The American Institute of Physics, 1993). Informative essays on Ptolemy, Copernicus, and Kepler. See also by Gingerich: *The Nature of Scientific Discovery: A Symposium Commemorating the 500th Anniversary of the Birth of Nicolaus Copernicus* (Washington, D.C.: Smithsonian Institution Press, 1975); *The Great Copernicus Chase and other adventures in astronomical history* (Cambridge, Massachusetts, and Cambridge, England: Sky Publishing Corporation and Cambridge University Press, 1992); *The Book Nobody Read: Chasing the Revolutions of Nicolaus Copernicus* (New York: Walker & Company, 2004).

Goldstein, Bernard, *Al-Birtuji: On the Principles of Astronomy* (New Haven: Yale University Press, 1971).

Grant, Edward, *Physical Science in the Middle Ages* (New York: John Wiley & Sons, 1971). Excellent introduction by a major scholar in this field.

Greene, Marjorie, *A Portrait of Aristotle* (London: Faber and Faber, 1963). An alternative interpretation of Aristotle's thinking, which characterizes his philosophy as governed by the interests of a biologist and has him continually analyzing and classifying, as if what were necessary to understand a subject was to divide it into categories

Heath, Thomas, *Aristarchus of Samos: The Ancient Copernicus. A History of Greek Astronomy to Aristarchus together with Aristarchus's Treatise on the Sizes and Distances of the Sun and Moon. A New Greek Text with Translation and Notes* (Oxford: Clarendon Press, 1913; New York: Dover Publications, 1981). Heath's 297-page history of ancient Greek astronomy introduces Aristarchus's text. Though now superseded by more recent and more accessible historical overviews, Heath's book usefully contains much material translated and quoted at length.

Hetherington, Norriss S., *Ancient Astronomy and Civilization* (Tucson, Arizona: Pachart Publishing House, 1987). Development of a science, and some of the cultural, historical, and intellectual interactions with civilizations that nurtured it. Now extended and superseded by Hetherington, *Planetary Motions* (Westport, Connecticut: Greenwood Publishing, 2006).

———, *Cosmology: Historical, Literary, Philosophical, Religious, and Scientific Perspectives* (New York and London: Garland Publishing, 1993). Less expensive paperback for classroom use with major entries from the *Encyclopedia of Cosmology,* including many from the Greeks through Copernicus and Newton.

———, *Encyclopedia of Cosmology: Historical, Philosophical, and Scientific Foundations of Modern Cosmology* (New York and London: Garland Publishing, 1993). Many major entries of over 5,000 words, including topics related to planetary astronomy from the Greeks through Copernicus and Newton.

———— , "Isaac Newton and Adam Smith: Intellectual Links Between Natural Science and Economics," in Paul Theerman and Adele F. Seeff, *Action and Reaction: Proceedings of a Symposium to Commemorate the Tercentenary of Newton's* Principia (Newark: University of Delaware Press, 1993), pp. 277–291.

———— , "Man, Society and the Universe," Mercury (November/December 1975), 1-4.

———— , "Plato's Place in the History of Greek Astronomy: Restoring both History and Science to the History of Science," *Journal of Astronomical History and Heritage,* 2 (1999), 87-110.

Heuer, Kenneth, *City of the Stargazers: the Rise and Fall of Ancient Alexandria* (New York: Charles Scribner's Sons, 1972). Elementary level.

Holton, Gerald, *The Scientific Imagination: Case Studies* (Cambridge: Cambridge University Press, 1978). See especially "Themata in Scientific Thought," pp. 3–24. Holton writes that a historian of science doing a thematic analysis is like a folklorist or anthropologist looking for and identifying recurring general themes in the preoccupations of individuals and of a society.

———— , *Thematic Origins of Scientific Thought: Kepler to Einstein* (Cambridge, Massachusetts: Harvard University Press, 1973). On themata in Kepler's work see "Johannes Kepler's Universe: Its Physics and Metaphysics," pp. 69–90; originally published in *American Journal of Physics, 24* (1956), 340–351. Holton is one of the most exciting writers on the history of science, and all his books are highly recommended.

Holton, Gerald and Stephen G. Brush, *Physics, the Human Adventure: From Copernicus to Einstein and Beyond* (New Brunswick, New Jersey: Rutgers University Press, 2001). Third edition of the 1952 *Introduction to Concepts and Theories in Physical Science.* Presents physical concepts clearly and in their cultural context. Arguably the best physics college textbook.

Hoskin, Michael, *The Cambridge Illustrated History of Astronomy* (Cambridge: Cambridge University Press, 1997). Outstanding text and illustrations from an outstanding historian of astronomy.

Jaeger, Werner, *Aristotle: Fundamentals of the History of his Development, second edition, translated by R. Robinson* (Oxford: Clarendon Press, 1948). The standard interpretation of Aristotle's thought: that he began very close to Plato's intellectual position and only gradually departed from it. Ambiguities in the dating of Aristotle's writings encourage such an analysis, because the resulting pattern then can be used to determine the chronological order of undated passages. For a different interpretation see Marjorie Greene, *A Portrait of Aristotle.*

Jones, Tom, *The Figure of the Earth* (Lawrence, Kansas: Coronado Press, 1967).

Koestler, Arthur, *The Sleepwalkers: A History of Man's Changing Vision of the Universe* (New York: Macmillan, 1959), A great read from one of the twentieth century's great writers. The part on Kepler is also published as *The Watershed* (Garden City, New York: Doubleday, 1960).

Koyré, Alexandre, *The Astronomical Revolution: Copernicus—Kepler—Borelli*, translated by R.E.W.. Maddison (New York: Cornell University Press, 1973). Especially rich in quotations illuminating Kepler's mentality.

———, *From the Closed World to the Infinite Universe* (Baltimore: the Johns Hopkins Press, 1957). Changes in man's conception of his universe and his place in it.

———, *Newtonian Studies* (London: Chapman & Hall, 1965). Essays analyzing how scientific ideas are related to philosophical thought and also determined by empirical controls.

Kuhn, Thomas S., *The Copernican Revolution: Planetary Astronomy in the Development of Western Thought* (New York: Random House, 1957; Cambridge MA: Harvard University Press, 1957). An outstanding analysis of relations between theory, observation, and belief in the development of Western astronomy to Copernicus, particularly within the framework of Aristotelian physics.

———, *The Structure of Scientific Revolutions* (Chicago: University of Chicago Press, 1962); second edition, enlarged (Chicago: University of Chicago Press, 1970). The most influential book ever written on how science works. The historian of science Steve Brush has asked, tongue only partially in cheek, if it is safe to expose students to Kuhn's ideas; will knowing that current scientific beliefs eventually will be overthrown and abandoned discourage students from studying current science?

Lindberg, David, *The Beginnings of Western Science: the European Scientific Tradition in Philosophical, Religious, and Institutional Context, 600* B.C.. *to 1450* (Chicago: University of Chicago Press, 1992). Outstanding introduction to the topic, highly recommended. See also Lindberg, ed., *Science in the Middle Ages* (Chicago: University of Chicago Press, 1978).

Lloyd, G.E.R., *Early Greek Science: Thales to Aristotle* (New York: W. W. Norton, 1970) and *Greek Science after Aristotle* (New York: W. W. Norton, 1973). These two slim volumes provide a good introduction to Greek science.

Lovejoy, Arthur, *The Great Chain of Being: A Study of the History of an Idea* (Cambridge, Massachusetts: Harvard University Press, 1936.) Traces the history of the idea of plenitude, that a good God created the universe full of all possible things, and continuity in nature.

Moore, Henry, *Writings and Conversations* (Berkeley: University of California Press, 2002).

Neugebauer, Otto, *Astronomy and History: Selected Essays* (New York: Springer-Verlag, 1983). See particularly "The History of Ancient Astronomy: Problems and Methods," pp. 33–98; reprinted from *Publications of the Astronomical Society of the Pacific*, 58 (1946), 17–43 and 104–142.

———, *Exact Sciences in Antiquity, second edition* (Providence, Rhode Island: Brown University Press, 1957; New York: Dover Publications, 1969). The major book on Babylonian astronomy.

———, *A History of Ancient Mathematical Astronomy* (New York: Springer-Verlag, 1975). The most detailed examination of technical details in Ptolemy's *Almagest*.

Newton, Robert R., *The Crime of Claudius Ptolemy* (Baltimore: The Johns Hopkins University Press, 1977). Argues that Ptolemy was the most successful fraud in the history of science. In Ptolemy's defense, see articles by Owen Gingerich: "Was Ptolemy a Fraud?," *Quarterly Journal of the Royal Astronomical Society, 21* (1980), 253–266, reprinted in Gingerich, *The Eye of Heaven: Ptolemy, Copernicus, Kepler* (New York: American Institute of Physics, 1993), pp. 55–73; "Ptolemy Revisited," *Quarterly Journal . . . , 22* (1981), 40–44, reprinted in *The Eye of Heaven . . . ,* pp. 74–80; "Ptolemy and the Maverick Motion of Mercury", *Sky and Telescope, 66* (1983), 11–13, reprinted in Gingerich, *The Great Copernicus Chase and Other Adventures in Astronomical History* (Cambridge, MA and Cambridge, 1992), pp. 31–35. For an overview of the first two decades of the debate, see Norriss S. Hetherington, "Ptolemy on Trial for Fraud: A Historiographic Review," *Astronomy & Geophysics, 38:2* (1997), 24–27.

Nicolson, Marjorie, *Science and Imagination* (Ithaca: Gold Seal Books, 1962). One of several studies by the author of the impact of science on the literary imagination. See also *Newton Demands the Muse: Newton's* Optics *and the Eighteenth Century Poets* (Princeton: Princeton University Press, 1946); *Voyages to the Moon* (New York: Macmillan Co., 1948).

North, John, *The Fontana History of Astronomy and Cosmology* (London: Fontana Press, 1994); also published as *The Norton History of Astronomy and Cosmology* (New York: W. W. Norton & Company, 1995). Perhaps the best single-volume history of astronomy, and authoritative. North's own scholarly research covers the time spectrum from Stonehenge through the Middle Ages to twentieth-century relativistic cosmology.

Pedersen, Olaf, *A Survey of the Almagest* (Denmark: Odense University Press, 1974; *Acta Historica Scientiarum Naturalium et Medicinalium, Edidit Bibliotheca Universitatis Haunienis, 30*). A thorough running commentary on technical matters in the *Almagest*.

Ragep, F. Jamil, *Nasir al-Din al-Tusi's Memoir on Astronomy (al-Tadhkira fi 'ilm al-hay'a). 2 volumes* (New York: Springer-Verlag, 1993). The major work on al-Tusi. See also Ragep's "The Two Versions of the Tusi Couple," in David King and George Saliba, *From Deferent to Equant: A Volume of Studies in the Ancient and Medieval Near East in Honor of E. S. Kennedy* (New York: New York Academy of Sciences, 1987), pp. 329–356.

Rochberg, Francesca, *The Heavenly Writing: Divination, Horoscopy, and Astronomy in Mesopotamian Culture* (Cambridge: Cambridge University Press, 2004). On changing interpretations of the significance of Babylonian astronomy in the history and philosophy of science, especially chapter one, "The Historiography of Mesopotamian Science." Anything by Rochberg is highly recommended.

Russell, Bertrand, *Philosophical Essays* (London: Longmans, Green, 1910).

Saliba, George, *A History of Arabic Astronomy: Planetary Theories during the Golden Age of Islam* (New York: New York University Press, 1994). A series of essays. See also "Whose Science is Arabic Science in Renaissance Europe?" and illustrations for al-Tusi's astronomy on Saliba's Web site: http://www.columbia.edu/~gas1/project/visions/case1/sci.3.html.

Small, Robert, *An Account of the Astronomical Discoveries of Kepler. A reprinting of the 1804 text with a foreword by William D. Stahlman* (Madison: University of Wisconsin Press, 1963). Chapter by chapter analysis with notes.

Sorabji, Richard, ed., *Aristotle Transformed: The Ancient Commentators and Their Influence* (London: Duckworth, 1990). See also his *Philoponus and the Rejection of Aristotelian Science* (Ithaca: Cornell University Press, 1987).

Stephenson, Bruce, *Kepler's Physical Astronomy* (New York: Springer-Verlag, 1987). Presents a detailed examination of Kepler's *Astronomia nova,* and argues persuasively that Kepler was guided by physical concerns to transcend traditional astronomy.

Stimson, Dorothy, *The Gradual Acceptance of the Copernican Theory of the Universe* (Gloucester, Massachusetts: Peter Smith, 1972). A pioneering effort from 1917 to trace changes in peoples' beliefs.

Swerdlow, Noel M., *The Babylonian Theory of the Planets* (Princeton: Princeton University Press, 1998). On Babylonian astronomy.

Swerdlow, Noel M., and O. Neugebauer, *Mathematical Astronomy in Copernicus's De revolutionibus* (New York: Springer-Verlag, 1984).

Taub, Liba Chaia, *Ptolemy's Universe: the Natural Philosophical and Ethical Foundations of Ptolemy's Astronomy* (Chicago: Open Court, 1993). A brief running commentary on *Book I* of the *Almagest* focusing on its place within Greek philosophical and scientific traditions and the degree of concurrence between Aristotle's and Ptolemy's ideas about philosophy and physics.

Thoren, Victor E., *The Lord of Uraniborg. A Biography of Tycho Brahe* (Cambridge: Cambridge University Press, 1990). The best book on both Brahe's life and his science.

Toomer, G.J., *Ptolemy's Almagest* (London: Duckworth, 1984; New York: Springer-Verlag, 1984). See also Toomer's encyclopedic entries "Apollonius of Perga," in C.C. Gillispie, ed., *Dictionary of Scientific Biography, vol. 1* (New York: Charles Scribner's Sons, 1970), pp. 179–193; and "Hipparchus," *ibid., vol. 15* (1978), pp. 207–234.

van Helden, Albert, *Sidereus Nuncius or The Sidereal Messenger Galileo Galilei. Translated with introduction, conclusion, and notes by Albert van Helden* (Chicago: University of Chicago Press, 1989). See also his *Measuring the Universe: Cosmic Dimensions from Aristarchus to Halley* (Chicago: University of Chicago Press, 1985).

Voelkel, James R., *The Composition of Kepler's* Astronomia Nova (Princeton: Princeton University Press, 2001). Interprets Kepler's *Astronomia nova* as a subtle argument rather than an undisciplined stream-of-consciousness narrative.

Westfall, Richard S., *The Construction of Modern Science: Mechanisms and Mechanics* (New York: John Wiley & Sons, Inc., 1971). On the new synthesis of science in the seventeenth century, from Kepler and Galileo to Newton.

————, *Never at Rest: A Biography of Isaac Newton* (Cambridge: Cambridge University Press, 1980). The most complete book on Newton and his science. An abbreviated and more accessible version is Westfall, *The Life of Isaac Newton* (Cambridge: Cambridge University Press, 1993).

INDEX

About the Author

NORRISS S. HETHERINGTON is the director of the Institute for the History of Astronomy and a Visiting Scholar with the Office of the History of Science and Technology at the University of California, Berkeley. He has written extensively on the history of astronomy and cosmology, and has edited *Encyclopedia of Cosmology: Historical, Philosophical, and Scientific Foundations of Modern Cosmology*, and *Cosmology: Historical, Literary, Philosophical, Religious, and Scientific Perspectives*.